Farewell, King Coal

The last shift

Farewell, King Coal

*From industrial triumph
to climatic disaster*

Anthony Seaton

DUNEDIN

EDINBURGH ◆ LONDON

Published by
Dunedin Academic Press Ltd
Hudson House, 8 Albany Street
Edinburgh EH1 3QB, Scotland

www.dunedinacademicpress.co.uk

ISBNs
9781780460772 (Hardback)
9781780465920 (ePub)
9781780465937 (Kindle)

British Library Cataloguing in Publication data
A catalogue record for this book is available from the British Library

Frontispiece: *The last shift*, sculpture by the author, photographed by Nick Seaton

Typeset by Makar Publishing Production, Edinburgh
Printed in Poland by Hussar Books

Contents

List of illustrations

List of Illustrations

Acknowledgements

My thanks are owed especially to three friends, Professor Bob Maynard, Professor Ken Donaldson, and Ms Sarah Lowry. Bob, who has an immense knowledge of air pollution, physiology, toxicology and English literature, read and commented on each chapter as I wrote it. Ken, toxicologist and musician, a colleague for 40 years, gave much helpful advice on the history and pathology of the dust diseases. Sarah, an oral historian at the London Royal College of Physicians who had recorded my own oral history, also kindly agreed to read and comment on each chapter especially for comprehensibility to the non-scientist. Their comments have been invaluable. Needless to say, no one other than I is responsible for any errors. I hope they are few.

I owe a special note of thanks to the Colt Foundation which from its inception has helped me and many of my younger research colleagues in Edinburgh and Aberdeen, and whose generous grant ensured the survival of the Institute of Occupational Medicine. I am also very grateful to the librarians of the Royal College of Physicians of Edinburgh, Iain Milne and Estela Dukan, and of the Institute of Occupational Medicine, Ken Dixon, for their assistance in tracing obscure references. More generally, I wish to record my indebtedness to colleagues and students throughout my career and to my own teachers, too numerous to name, but who are remembered with gratitude by me. Finally, so many thanks to my dear wife Jill. She heard me say 'never again' when I finished my last book 18 years ago, yet has tolerated with good humour and support my spending much of my 80th year breaking that vow.

Foreword

Farewell, King Coal is an evocative title for Professor Seaton's history of coal and the consequences of its extraction and combustion on the health of the planet and its inhabitants. In Professor Seaton, coal has a biographer who writes with authority, clarity and often personal knowledge both of the scientific advances and of the scientists whose research has provided our current knowledge of the nature and impact of these effects.

His history of coal is informed by his experience as a respiratory physician and medical scientist, working in the field of occupational lung disease, often directly involved in important scientific advances. He interweaves this experience with the history of the increasing understanding of the risks which coal and its combustion pose. Following a period working in West Virginia investigating the effects of inhaled coal dust on the lungs of Appalachian miners, he returned to the UK as chest physician in Cardiff, working in close proximity to the Medical Research Council Pneumoconiosis Research Unit (PRU) where much fundamental research on the risks of inhaled coal dust to miners was undertaken. The work of the PRU in the 1940's and 1950's provided understanding of the nature of coal workers pneumoconiosis (CWP), of the factors leading to the development of the associated disabling Progressive Massive Fibrosis (PMF) and, critically, of the means to prevent it.

He describes the outstanding leaders of this remarkable endeavour. Two figures stand out: Jethro Gough who demonstrated that PMF in coal miners need not require silica exposure, but could be a consequence of exposure to coal dust alone; and Archie Cochrane who showed that the risk of PMF increased directly with increasing category of simple pneumoconiosis, itself a reflection of the quantity of coal dust retained in the lungs, providing one of the means to prevent PMF in coal miners. Cochrane, who was later to achieve fame for his monograph, "Effectiveness and Efficiency", maintained that his work on CWP was his most important contribution to scientific research.

Professor Seaton subsequently moved to Edinburgh to lead the Institute of Occupational Medicine (IOM). Initially set up by the National Coal Board, IOM was responsible for the regular surveillance of coal miners by questionnaire, lung function tests and chest radiograph, together with measurements of the levels of coal dust in mines. The results of this remarkable programme of work, undertaken over many years, allowed estimation of an exposure-response relationship for the concentration of airborne coal dust and the risk of CWP, providing a rational basis for an exposure limit for underground levels of airborne coal dust. It also enabled a series of publications which demonstrated unequivocally the increased risk of chronic obstructive pulmonary disease (COPD) in coal miners, both cigarette smokers and non-smokers, providing resolution to a previously hotly disputed topic.

The UK Clean Air Act in 1956 followed the 1952 London smog, which was responsible for some 4,000 excess deaths. By enabling the prevention of air pollution episodes the Clean Air Act was for many years considered to have solved the problem of air pollution in the UK. However, research, particularly in the USA in the 1990's, showed that air pollutants at ambient levels, particularly small particles, both from coal combustion and motor vehicle exhaust, were a continuing source of ill health and premature death. Dr Robert Maynard persuaded the Department of Health to set up a committee to make recommendations about standards for outdoor air pollutants – the Expert Panel on Air Quality Standards (EPAQS) - which he invited Anthony Seaton to chair. Through this he became very familiar with the science of air pollution and the risks it posed to human health. Counter-intuitively the excess deaths associated with particulate air pollution were more from cardiovascular than respiratory causes. With his colleagues in Edinburgh, Professor Seaton suggested this was the consequence of the inflammatory reaction in the lung, provoked by the inhaled particles, causing an increase in blood coagulability. Subsequently known as the 'Seaton hypothesis', this has proved to be the catalyst for much subsequent research. More recently he has extended his interest to climate change and of the role of King Coal – and King Oil – in causing this. In this book he provides the clearest exposition of the science of climate change I have read.

Professor Seaton bids an eloquent farewell to coal, describing its trajectory from the primary energy source driving the Industrial Revolution to current recognition of the role of fossil fuels, particularly coal and oil, in air pollution and climate change. However, while appreciation of these effects in many parts of the world is stimulating the increasing use of substitutes for fossil fuels, worldwide coal remains the fuel most used for electricity generation, currently accounting for nearly 40% in total and for more than three quarters of electricity generation in India.

Professor Seaton's history is timely, relevant and of universal interest. We should be grateful to him that he decided to spend his 80th year informing us with such clarity and verve.

Professor Sir Anthony Newman Taylor
CBE, FRCP, FFOM, FMedSci
Professor of Occupational and Environmental Medicine
National Heart and Lung Institute
Imperial College London

Introduction

For God's sake, let us sit upon the ground
And tell sad stories of the death of kings…

My generation grew up in the age of coal. It heated our houses, powered our factories and was the source of the gas we cooked with and of the light in our houses and streets. As children in the 1940s we were fascinated by steam engines and longed for a model railway set – I finally bought one when I had sons of my own. Leaning over the railway bridge parapet we laughed as we were enveloped by the shower of soot from the engine; sometimes we trespassed on the lines to put a halfpenny on them so the train would squash it into a penny. We became familiar with the smoke in our cities from houses, factories and the railways, and experienced the dense acrid winter fogs that people called smog. For a while during the 1939–45 war, smoke machines in the park across the street belched out dense black clouds in a vain attempt to obscure the city of Liverpool from the German bombers, the idea of a smoke screen having been derived from the ones put out by warships in the First World War to hinder the view from the enemy's vessels. We boys ran through this smoke, challenging each other to see who could last longest without falling into a fit of coughing. In Liverpool, we saw the fleets of coal-powered cargo ships preparing for their hazardous journeys – west across the Atlantic to America to bring us supplies, and north round Scandinavia to take aid to the Russians after they were invaded by Germany and became our allies.

When my father went abroad with the army, we moved to rural Yorkshire to stay with our grandmother and avoid the bombing which had been such a feature of our early childhood. I watched her cooking in an oven over a coal fire and ironing the clothes with a flat iron heated over the same fire in the kitchen. Later I learnt how to light the fire in the living room, using newspaper to light the sticks used as tinder, blowing in the sputtering flames and drawing the fire by holding the paper across the top of the grate. I saw the grey ash left behind in the morning as my mother cleared the grate before going to the coal shed to fill the scuttle by the fireplace, and noticed that some coals left more ash than others. We shivered in our house until the afternoon, when the fire was lit and we gathered round it to keep warm in the winter evenings; our home made its own contribution to the smogs.

When I was in my teens we had to stop burning coal and obtained what was called smokeless fuel, compressed cobbles of powdered coke that produced rather less smoke and very little ash. A law had been passed to reduce the air pollution, which we had, until then, thought of as one of the normal circumstances of city life; and the air did indeed become cleaner. Later, more efficient oil-burning stoves and gas central heating were introduced. Factories and power stations moved out of the cities and the winter smogs no longer occurred. By 1956 I had become a medical student and saw how many people came into hospital with coughs and breathlessness, learning of the disease that was called chronic bronchitis and was said to be due to cigarettes and air pollution; but I did not realise then what a part that disease and coal were to play in my future life as a doctor and medical researcher.

From the 1950s petrol was no longer rationed and became widely available. Cars were becoming more frequent in the streets, the steam engines were giving way to much less interesting diesel locomotives, and coal usage was increasingly confined to power stations in the country, with tall chimney stacks that often exported the smoke to other countries in northern Europe. We heard of acid rain falling there, damaging the environment of rivers and lakes. Coal nevertheless remained the dominant fuel in Britain through the nineteenth and twentieth centuries, over which period the effects of its mining and use on both the workers and the general population were increasingly recognised.

After qualifying in medicine and learning my trade in Liverpool and Stoke-on-Trent, from 1969 to 1971 I spent time in West Virginia, USA. This was a coal-mining state and we saw long chains of coal barges snaking north to Pittsburgh on the Monongahala and Ohio rivers. It was there that I started research on miners' lung diseases before moving to work as a chest physician in South Wales, spending part of my time researching asthma. In 1978 I moved to Edinburgh and reversed these roles, researching occupational diseases and working part-time as a chest physician. In Britain, by then, oil had replaced coal as a fuel in many applications. Cheaper, more accessible coal was available from elsewhere, and the British mines had started closing; the final phase of the coal industry had commenced. Oil and nuclear were considered the fuels of the future and motor vehicles rather than houses and factories became the dominant source of pollution in our cities. My final move took me to Aberdeen, the centre of the North Sea oil industry, where I was able to start researching the health effects of air pollution and, on retirement in 2003, I went back to Edinburgh and investigated the story of climate change.

Now, as an old man, I look back on the decline and death of the coal industry with mixed feelings and say, echoing the words of Shakespeare's *Richard II*, 'Farewell King Coal'. But I watch with interest the potential decline of oil as a fuel, soon necessarily to be followed by gas, a switch away from fossil fuels driven by understanding of climate change. This is my personal obituary of coal in the context of an individual's medical career and a population's increasing understanding of mankind's place in the ecology of the Earth. It is the story of the most disruptive technology ever introduced by mankind, and the consequential increasing prosperity of the western world, but also of the deaths and diseases caused by coal, its mining, utilisation and combustion, and of the scientific disputes that surrounded the medical discoveries. As such, it is an important part of the story of mankind's unending struggle to survive on this restless planet in harmony with the animals, microbes, and plants that share it with us:

> *for within the hollow crown*
> *That rounds the mortal temples of a king*
> *Keeps Death his court, and there the antic sits …*

The metaphor is apt, as we shall see.

Figure 1.1 One of many erratic stones by the River Almond, deposited when the ice cap melted.

Chapter 1

Early beginnings

Close to my home on the western outskirts of Edinburgh is a small river, the Almond. As you walk beside this river you see large stones of great age, deposited some 12,000 years ago when the ice was melting at the end of the last Ice Age. They are called erratic stones, from the Latin *errare*, to wander. They differ in mineral constituents from the rocks where they are found, showing that they have been carried often great distances in the past. They played an important part in the early discovery of the ice ages, when it became apparent that in the past, Alpine glaciers had been much more extensive. This is discussed further in chapter 11 (see Fig.1.1). Walking towards the river's mouth you may see some places where erosion has revealed deposits of shale and even coal. You pass by the ruins of old mills, dating back to the fifteenth century, that used the flow of water to drive machines that first worked grain for bread, then iron to make nails and agricultural tools. As the Almond flows into the Firth of Forth you come to the village of Cramond; excavations here have revealed the site of a Mesolithic (Middle Stone Age) settlement dating from 8000BCE, a relic of the first hunter-gatherer people who, when the sea level was much lower, had migrated across Doggerland from mainland Europe to inhabit what became known as Scotland. In Cramond you can see the site of the fort that the Romans built around 100CE to supply their soldiers on the Antonine Wall to the west, stretching across central Scotland as far as Glasgow, defending their Empire from the northern tribes.

The ruined iron mills point to the use of water power, a major source of energy in early industrial times (see Fig. 1.2). As you stand by the Firth of Forth at Cramond, looking across the river towards Fife or to your right, towards the mouth of the river along the shore of East Lothian, you see land that has for centuries been exploited for coal, together with a huge power station that burned the fuel. To your left, beyond the three Forth Bridges, is more coal country on either side of the river and another power station. Both are now closed as we have progressively reduced our dependence on fossil fuels. Beyond Barnbogle castle on the estate of the Earl of Rosebery lie the lands in West Lothian that were exploited for mining oil shale, the world's first commercial source of mineral oil, discovered in 1850. Should you go a few miles further to the west, to the start of the Romans' Antonine Wall, you can see the canals built to connect the ports of

Figure 1.2 The mill pool at Cramond Brig that once supplied one of the 18th-century iron mills.

Glasgow, Falkirk and Edinburgh, from the Irish Sea to the North Sea, which transported coal from the local mines to the two major Scottish cities. These reminders of industry are now just that – reminders. All around is attractive scenery – fields, woods, hills and distant mountains – but where I live you can't ignore history, the migrations of mankind, and the evidence of our search for energy.

The origins of coal

The story of coal starts long before the last Ice Age or the first steps taken by *Homo*, the human genus, but it is nevertheless a story intimately bound up with Man's history, his success as the Earth's dominant species and, perhaps, his ultimate decline. It starts in what is known as the Carboniferous or coal-bearing period, about 300–350 million years ago, part of the Palaeozoic (ancient animals) era, when the continental land masses were moving together to form one massive continent, Pangaea, and volcanic activity was at a peak. Much of the Earth's surface was ocean, a large area of ice formed at the South Pole, and most of the land was marshy and covered with vegetation, especially tropical ferns, huge primitive pines, and mosses. An extinction of many animals and plants had occurred over the previous few million years (the reasons are still being debated but a decline in oxygen in the atmosphere may have been responsible) and during the Carboniferous many new animal forms evolved, notably species of fishes, insects and reptiles.

The age of the amniotic egg, an evolutionary device to protect the developing embryo from adverse environmental conditions, had arrived, allowing reptiles to colonise the land. With this, the age we know best for the dinosaurs was on its way.

Fluctuations in sea level as the mainly southern ice formed and melted caused periodic flooding and death of plants but, probably because fungi and bacteria had not yet evolved to the point of being able to break down tough plant material, much of the submerged organic matter did not decay but became peat, as may still happen today when plants die and are deposited in conditions of low soil oxygen levels. Ultimately this peat, buried and subjected to high pressures and temperatures, was fossilised and converted into coal. As the moving continents collided, huge pressures built up, the land folded, and tall mountain chains formed. In some places the pressure was particularly high and the compression of the coal resulted in an especially hard form called anthracite. In some parts of the landmass the layers of coal were laid down under sea, and in others under fresh water, so that coal seams in some places are found in limestone (the fossilised relics of marine animals) and in others in quartz-bearing rock such as sandstone or millstone grit. Miners the world over recognise the fossil plants in coal and the presence of seams or intrusions of other minerals when they are cutting coal.

This period of the Earth's turbulence left us the mountains, oceans and continental landmasses with which we are familiar today. Now earthquakes, tsunamis and volcanic eruptions remind us on a human timescale that the motion of the Earth's surface, continental drift, continues. We shall return to this when we discuss climate change in chapter 11. The layers of coal had been distributed around the planet and lay undisturbed for millions of years, until mankind arrived on the scene.

The origins of mankind

More than two and a half million years ago in East Africa, the first humans of the genus *Homo* had already evolved from large-brained apes that were able to walk on their hind legs and use tools.[1] Some two million years ago they spread over much of what are now Europe, North Africa and Asia. Over millennia, driven by the evolutionary pressures of their environments, *Homo* evolved into multiple different species, of which a European one became known as *Homo neanderthalensis,* or Neanderthal man. This species and another more ancient Asian one called *Homo erectus* were well enough adapted to survive until the last Ice Age, a total of over two million years – to date the most successful of all the human species. Over this long period, starting around 300,000 years ago, humans had discovered the use of fire and learned how to cook food, and about 150,000 years ago one large-brained species became dominant in Africa. About 80,000 years ago it migrated across the Red Sea to populate the world from Western Europe to Asia and Australia, and ultimately across the Bering Straits to the Americas. We have given it a name reflecting its (and our) apparent relative intelligence; *Homo sapiens* or wise man had arrived and, unlike the other species of *Homo,* survived the last Ice Age in multiple refuges that would have had relatively benign climates. With some admixture with the genes of other *Homo* species, two or three of those early people who migrated out of Africa are the ancestors of everyone on Earth today. Genetic evidence shows that we all have a single common grandmother from Africa.[2]

Many features contributed to the survival success of *H. sapiens;* a large, well-connected brain enabled learning from experience, ability to make and use tools, to construct shelter from the elements, to use fire to cook and to clear land to grow crops, and to deter and out-think predators. Over some 80,000 years we have evolved in our different environments to look and behave very differently, but in all cases to be beautifully adapted to that environment, be it the tropical forest, the Australian desert, the savannah, the Arctic wastes or even the towns and cities that over time evolved to accommodate inter-dependent communities, and that characterise what we think of as 'civilisation'. In the course of this evolutionary passage, we discovered a portable fuel that gave more heat than wood, peat, or even charcoal – our ancestors had found coal.

Exploitation of coal

At first this fuel had very limited value, but archaeological evidence of its use nevertheless dates back 5000 years in China, and the ancient Greeks recorded using it in metal working. The Romans used it also for heating in the cold northeast of Britain up to around 400 CE. Coastal trading and even export to continental Europe had started as early as this time. Its use in lime burning and metal working was rediscovered in the northeast of England in the thirteenth century, and from the earliest years of its industrial use there are records of complaints about the smoke.[3] As a fuel it was competing with wood, peat and charcoal; it was recognised to generate more smoke than these, although it usually produced more heat. In Anglo-Saxon and Norman times only castles had chimneys, and domestic heating was from a wood or peat fire in a pit in the floor of the hut, the smoke escaping through the thatch roof or from a hole in it, as is seen today in such places as Papua New Guinea or rural India. In England, laws were enacted to have these fires covered by a lid at night time, when a bell would be rung in the village – in Norman French *couvrez feu,* from which we derive *curfew.* From the fifteenth century, wood became increasingly scarce in Britain from its use in the newly introduced cast iron industry (the iron was used principally for making weapons) and in building ships and houses, and coal started to be used for domestic heating; this era saw the introduction of the iron grate and the once familiar poker and tongs in some houses, derived from their prior use in blacksmithing trades. However, by the fourteenth century, in London coastal trade had already brought enough coal for small industries to use it extensively, and the first smoky fogs occurred. In 1306 King Edward I of England issued a proclamation forbidding artificers to use 'sea coal' in their furnaces – the term probably derived from its importation by sea to differentiate it from charcoal – and the city of Newcastle in the northeast of England became the centre of a thriving trade in coal, even exporting it to France and the Low Countries.[4] This valuable cargo required protection by a special fleet of Admiralty ships against attack by other nations. Since those times, in Britain, 'taking coals to Newcastle' has been a commonly used metaphor for unnecessary trade or advice.

Until the fourteenth century, the use of coal had largely been confined to areas where outcrops, especially on the coast and river banks, made it easily accessible (and provided an alternative derivation for 'sea coal'). From then on, the trade flourished and by the seventeenth century a large fleet of hundreds of ships carried coal from Newcastle to European and English ports, especially London. The seventeenth century also saw great development of the use of coal in metal manufacture, especially of iron. By the end of that century a method had been developed of making a less smoky and more efficient fuel by cooking (or carking) the coal to make coke, by driving off the volatile matter. The eighteenth century saw the development of canals, primarily to transport coal to the developing cities of the industrial era, spreading the pollution from coal burning more widely. And finally, in the nineteenth century the railways took over this trade, surviving in Britain up to the late twentieth century.

Mining

Mining, the process of digging into the Earth's surface to exploit minerals, is an ancient trade. Indeed, the metals mined and exploited by early people define the era; sequentially the Stone, Bronze and Iron Ages, named after the predominant artefacts found by archaeologists. Mining would have started by digging pits to remove surface layers of earth and stone, and open-pit or strip mining is still used widely to produce coal. In Britain, the oldest known underground mines are those dug from about 3000 BCE by Neolithic people near Brandon in Suffolk, to obtain flints for arrow heads and axes. These mines, known locally as Grime's Graves, are pits dug down a few metres to the limestone seam in which the flints are found; the miners then used deer antlers as picks to extend the mine out radially. Such techniques, but using iron picks, were also employed in some of the first underground coal mines. More commonly, however, early coal mines exploited surface outcrops of coal seams, digging along the seam into the Earth's surface layers – the so-called drift mine. This remains a common method of mining to this day, and was inherited from ancient metal mining; I was surprised to find such very simple methods used in smallholdings by individuals in West Virginia USA when I lived there in the late 1960s. When they needed coal they went outside and dug some up.

The first metal mining in Britain was for copper and tin, which were easily worked and smelted to produce bronze – hence the Bronze Age, from about 2500 to 700 BCE. These mines, for example for tin in Cornwall, copper in North Wales and the Isle of Man, and iron in the Lake District, continued to be exploited into the twentieth century. Worldwide, iron began to replace copper as the mineral of choice for manufacture and weapons from about 1200 BCE, and the Iron Age reached Britain around 600 BCE, so mining was well established in these islands by the time the Romans first invaded in 55 BCE. However, the Romans brought with them more advanced mining techniques, which have been investigated particularly in the Welsh gold mines in Dollaucothi in Carmarthenshire, and were described by Pliny the Elder in his book *De Re Naturae* (on natural history) before he died at the site of the eruption of Vesuvius. These included hydraulic methods to wash away unwanted earth, the use of fire to crack rock, and tunnelling, so that by the time coal started to be exploited on a grand scale most of the basic mining techniques were present in Britain.

Trade, industry and the Industrial Revolution

The fact that we have named the early eras after the predominant minerals used indicates the importance of technological innovation in the history of human civilisation. Stone, bronze and iron: each characterised a step-change in human development and each stimulated trade, exchange of commodities and manufactured goods between people, tribes and nations. Conflicts, leading to warfare and alliances, arose from the need for commodities; the silk trade, spices, precious metals as luxury items, but more basically, the need for water, energy and food crops for survival and protection against the elements.[5] The need of the growing population of humans to develop technologies and to trade has led to the rise and fall of empires, something we Celts and Picts in the British islands first learnt when the Romans invaded. Our island status has hindered conquest since then, but both the Norse Vikings from the eighth century and the Norman French in the eleventh century successfully invaded us and have made important contributions to who we are, our genes, our language and our culture.[6] When Shakespeare in *Richard II* wrote the well-known soliloquy quoted in the introduction, he was using words derived from three invasions: Roman, Norse and Norman French. Being an island has also meant that we have been required to develop trade by sea from the very earliest times, as well as continuing to develop our own indigenous sources of food and commodities. This first became obvious to me as a child, when blockade of our trade routes across the Atlantic by Nazi Germany's U-boats led to severe shortages of food; rationing accompanied by long queues outside shops became part of everyday life. We shall revisit this in chapter 13.

This last observation, on blockade by submarine, points to another very frequent accompaniment of sea trade, naval warfare. The ready availability of wood and copper allowed the development of a powerful English navy in the fifteenth and sixteenth centuries, a time when Portuguese and Spanish explorers were opening up trade routes to the Far East and the Americas. Starting with legitimised piracy, English naval fleets gradually established dominance over our European rivals, the French, Spanish and Portuguese, and established trade routes and colonies around the world. This of course involved the exploitation of the lands and the people who were colonised, even including the transportation of humans to labour as slaves, a trade which was only finally abolished in the nineteenth century, but which sadly continues in a less open manner to this day in some parts of the world.

In 1707 England and Scotland, already having had a sovereign in common for a century, united politically to form Great Britain. This new country had many colonies in the Americas, Africa, India and the Far East, as did other European countries. The wealth derived from this exploitation drove improvements in education and technology across Europe but also encouraged continued warfare between European countries. To win wars requires technological advancement, and this technology may often be transferred to peaceful purposes. While people were rising in revolt against their leaders and at war with their neighbours, an industrial revolution was beginning in Europe.

Two of the great universities of the newly named Great Britain were in Glasgow and Edinburgh. At the end of the eighteenth century England had only two universities, Oxford and Cambridge, whereas Scotland had six – two each in Aberdeen and Glasgow, and one each in St Andrews and Edinburgh. Glasgow University, the second of these, had been founded in 1451 by William Turnbull, Bishop of Glasgow, who had studied at the older St Andrews and Padua universities. Edinburgh University, the youngest of the six, had been established in 1583 by Royal Charter of King James VI, later to accede to the British crown as James I. The union of the two countries in 1707 under James brought to Great Britain not only the enhanced educational opportunities previously confined to the Scots, but also the important links that already existed between Scottish and European universities.

The eighteenth century, the age of the Scottish enlightenment, was a period in history when intelligent children from all walks of life had the possibility of acquiring an education and entering a profession. Three notable beneficiaries of this period were James Hutton, Adam Smith and Joseph Black. All knew each other and became friends; all three made seminal contributions to the tale I am telling. Hutton (1726–97) was an Edinburgh graduate and landowner who, as a consequence of his interest in the productivity of his land, observed the different strata in the rocks, deducing from this the great age of the Earth and the constant movement of its surface, thus leading to the science of geology and the discoveries that underpin plate tectonics. Adam Smith (1723–90), professor of moral philosophy in Glasgow University, in his books, *The Theory of Moral Sentiments* and *The Wealth of Nations,* started the great debate that continues to this day and became known as economics. Joseph Black (1728–99), who was professor of medicine in Glasgow and later Edinburgh universities, established quantitative methods in chemistry and discovered latent heat and the first gas, carbon dioxide.

In 1776 the British American colonies, much extended in area after defeat of the French in the Seven Years War, started the rebellion against the British crown that was to become the American War of Independence. Adam Smith's book, *The Wealth of Nations,* was published that same year. Its theories about the 'invisible hand of the market' and free trade had an immediate impact across Europe. In it he proposed that ideal markets, driven by self-interest and controlled by competition, would regulate themselves. He theorised that the beneficiary of the wealth generated would be society itself. Coincidentally, that same year a patent was taken out for a new improved steam engine that ultimately made available to society powerful engines for pumping water, transporting goods and people, and running the mills and engines of manufactories, or factories as they became known. James Watt (1736–1819) had been working as an instrument maker in Glasgow University when he was asked to repair one of the Newcomen engines currently in use and, noting its inefficiency, embarked on a prolonged series of experiments that resulted in his designing a condenser and steam jacket to conserve the thermal energy of the system. In doing this, he had independently arrived at the concept of latent heat,[7] and he was later to become friends with its discoverer, James Black, as well as with Adam Smith. It is not unreasonable to think that the Industrial Revolution, which was just starting with the factory system of production, owed its major impetus to the ideas of these three contemporary Scottish philosophers.

The success of the factory system hinged on increased productivity from the introduction of powered machines and increasing specialisation of the workforce, as advocated by Smith. The power was obtained originally from water, so factories had been situated by rivers. Engines allowed both greater, more reliable power and also more flexibility in where to site the factory. The more efficient they were, the greater the profitability of the concern. In mines, steam engines were to transform the means of ventilation, the pumping out of intruding water, and the transport of materials and people, and of the product, coal.

By the end of the eighteenth century, with France wracked by revolution, Britain was the dominant world power in spite of its loss of the American colonies. The French had beheaded Louis XVI in 1793, Napoleon had been finally defeated at Waterloo in 1815, and Victoria became Queen in 1837; Britain could claim to rule the waves, but coal had become King. Sea power was fundamental to this, including both the Royal Navy and the merchant fleet. Steam power was introduced for ships in the 1820s and the screw propeller, more efficient than the original paddle wheels, was first used in 1839 on the ship *Archimedes*. This increased the need for coal and miners. Not only men (who did the hewing and drilling) but also children and women were employed, and in the search for energy the number of mines in Britain increased. By 1901 coal production in Britain was estimated to be 225 million tons annually and 780,000 men were employed in the UK's mines. But in that year of Queen Victoria's death, over 1100 miners gave their own lives while working in Britain's coal mines.

Chapter 2

Earth, air, fire, and water: the dangerous life of the coal miner

For those of us who live and work on the Earth's surface, a visit to a coal mine is a memorable experience. Taking off all your clothes save underpants and socks in the locker room at the pit-head (it's going to be hot down there), you put on overalls and safety boots and collect your helmet. You get a lamp for your helmet and a battery pack for your belt. With the lamp comes a numbered tally. You collect your self-rescuer, a portable oxygen generator for emergency use, fixing it to your belt. On the way to the cage you notice the piles of discarded cigarette butts, the remains of the last drags on them as the miners prepared for a shift without a smoke. Lots of men crowd in with you and the descent starts slowly at first but very quickly speeds up; your stomach comes into your mouth and your ears pop as you feel yourself being lifted as it descends very rapidly to pit bottom. The men talk and joke in the local dialect, using what is euphemistically called 'pit language'. There is an obvious exclusive camaraderie among those who work in the dark below ground, from which you are excluded: this is a job for special people. Most of them are following their fathers and grandfathers down the pit – the job is what defines them. You reflect that this is the traditional, feudal way in which society had been organised; the men have their defined roles and pass them to their sons, and you remember that this was true also of yourself in your profession.

At pit bottom you are surprised to find the mine is quite spacious at first. You switch on your lamp and start walking along a long passage or roadway, the inbye, supported by steel arches at first, following the overman who is accompanying you and who carries a Davy lamp and a stick which denotes his rank, but may also be used to lift his safety lamp to the roof to test for methane. You feel the wind behind your back. After a walk (or if the roadway to the face is a very long one, a ride on a conveyer belt or in a tub) it becomes pitch black apart from the light from your helmet lamps. You notice various dark branch tunnels leading to other faces and reflect on the need for that tally in case you get lost, though for miners it comes into its own in case of an explosion or fire. Gradually the roof gets lower and you bump your head, learning one reason for your helmet; you continue on your way in a stooping posture, difficult for those not used to it. It is hot and noisy, a conveyor belt running alongside you. You see the dust in the air, caught in the beam of your light, but you notice with surprise that the walls are grey-white. The mine has been dusted with limestone to reduce the risk of explosion.

As you reach the coal face the roof gets very low and you walk bent over; the noise reaches a crescendo and you feel a draught of air as you turn a right angle into the low, pitch-dark tunnel where the noise is coming from. You may have to crawl at this point, as the coal seam is quite narrow, and you are glad of the knee pads you were provided with. The machine comes towards you, its cutters revolving and slicing coal from the seam, which falls onto a belt and is carried away (see Fig. 2.1). After the coal cutter has passed you, the roof above the space where the coal was is held up by horizontal props that are moved forward by another machine, to protect the miners who will be working there when the coal cutter returns in the opposite direction. Behind you, the roof that was supported a moment ago by this machine falls in a pile of rubble. Over a distance of several hundred metres, the coal face gradually advances into the seam. The air is hot, and damp with the water that is sprayed continuously onto the cutter and onto the falling rock to reduce the dust concentration in the air. You see the miners, black faces emphasising the white of their eyes and teeth, operating the machines. At the end of the face, perhaps 200 metres long, you make another right angle to walk back with the flow of the air to return in the outbye to the pit bottom. As you make your way out, not far from the face, you may see an instrument hanging in the tunnel leading from the face,

Figure 2.1 Coal miner operating shearer on long wall face. (Source: Scottish National Mining Museum.)

sampling the dust in the air that the miners have been breathing. You return to the surface in the cage and hand in your tally and equipment. In the shower at the pit head you wash off the black coal dust, noticing how much has accumulated even after such a short visit.

This was a description of a modern British mine in the 1980s, towards the end of the story of mining in this country, and it illustrates some of the ways in which the dangerous trade had gradually been made safer over the 400 year history of deep coal mining in Britain. The ancient Greek philosopher, Aristotle, spoke of the four elements, earth, air, fire and water, which constituted everything, the environment in which we live. The miner sees danger in all of these – the earth may fall on his head, the air may poison him, the mine gases may ignite and explode, and an in-rush of water may drown him. Mining is a trade in which, uniquely, all four elements constantly threaten the lives of those engaged in it.

The equipment of the mine and the miner, such as battery-powered lamps, safety boots and helmets, the roof supports, the ventilation to remove gases and reduce dust concentrations, water to suppress dust, a self-rescuer to provide oxygen in case of a gas or fire emergency, and the training and supervision that miners receive, are all intended to afford some protection. Each man looks out both for himself and his workmates. However, such well-run mines could not be said to have been typical of mining conditions everywhere, even in the 1970s and 1980s. Nor does this account give a picture of the work of men who drive the roadways to the coal face using drills and explosives to break the rock. I was able to get a taste of much more primitive conditions in Indian and even some small United States mines, where men still worked lying on their sides cutting coal with pickaxes and digging it out with shovels, where as you walked to the face you heard ominous creaking noises from the rock strata above your head. In an Indian mine, in which the miners all worked using hand tools to cut the coal, I even saw a sad-looking caged canary, its yellow feathers obscured by coal dust, being used as a detector of carbon monoxide. In such conditions it is easy to visualise the dangers that lurk underground, and even to understand why superstitions about evil spirits, goblins and dangerous miasmas arose in mining communities in the past.

Early coal mining [1]

The first places in Britain where coal was extracted were by the sea in the northeast – Northumberland and around Edinburgh and the Firth of Forth. There are records of rights being awarded to the monks of Holyrood Abbey in Edinburgh to extract coal in the late 1100s and of the trade of this coal after transport by sea to London in the early 1200s. Similar rights were awarded to monks in Newcastle about this time. From the fourteenth century the simple quarries and holes from which coal was extracted were succeeded by the type described as pit and adit, similar to the Neolithic mines mentioned in chapter 1, with a shaft and a horizontal underground gallery. The coal was hauled up by a windlass operated manually or by horsepower. The adit was a passage into the mine, often used with a water gate to drain the workings.

It was from this that the nascent mining industry confronted its first major problems: the risks from rock fall and from the ingress of water into mines. The deeper the mine, the greater the problem, and towards the end of the fifteenth century records show the introduction of pumps by monks in mines in Northumberland. The next great problem specific to coal mining was fire, spontaneous and prolonged combustion of coal having been recorded through the sixteenth century. Increasing depth was also associated with poorer ventilation and accumulation of gases, a problem addressed in the first great book on metal mining and refining, *De Re Metallica*, by Georgius Agricola and published in Basel in 1556. Agricola was born Georg Bauer in Saxony in 1494 and, after a spell as a teacher of Greek and Latin, studied science and medicine and became town physician in Joachimstal, now Jakimov in the Czech Republic, centre of the metal mining industry. The principal product was silver, which was used to make the coin called a joachimstaler or taler, later providing the etymology for the word dollar.

Agricola was a prolific writer and *De Re Metallica*, his most famous book, translated from the Latin into English by Herbert Hoover and his wife Lou,[2] reflects his many investigations into the conditions and mechanics of mining. He gives a detailed description of the technology available in Europe in that era, both for mining and for prevention of the hazards. It describes the harmfulness of dust, *eating away the lungs to form a terrible consumption*, and the dangers of both suffocation from stagnant air and death from inhaling fumes from fire. He goes on to describe methods of ventilation, of removing water, and of transportation from the mines, as well as the extraction and refining of the metals. His illustrations show a technology that was to change little until the era of steam some 200 years later (Fig. 2.2).

About the same time, in Britain, an early comment on the dangers of mining came from Dr Keys, co-founder of Gonville and Caius College in Cambridge. In referring to poisonous gases from fire in coal mines, he wrote: *the unwholesome vapour whereof is so pernicious to the hired labourers that it would immediately destroy them if they did not get out of the way as soon as the flame of their lamps becomes blue and is consumed.*

Developing Technology

British coal mining remained quite primitive through most of its history up to the late nineteenth century.[3,4] Coal was obtained by pick and shovel and carried out of

Figure 2.2 Agricola's illustration of a 16th-century metal mine.

the mine on the backs of women or ponies. Women and boys were also employed to drag coal trucks and to operate doors to direct ventilation. In Scotland, particularly, the condition of the miners was little different from slavery, they and their families having been bound to the owners and sold with the mine right up to the start of the nineteenth century. As mining engineers developed the industry in order to increase productivity in a competitive market, so increasing hazards to the workers would become apparent. Behind this short review of early mining developments lies a story of increasing risks to miners of death and injury from falls, floods and fire as mines became more complex and deeper.

The main early method of coal mining in Britain was referred to as room and pillar, whereby coal was cut from the seam leaving uncut pillars to support the roof, the relative size of these being determined empirically. Large rocks were originally removed by being heated with fire, then cracked with cold water, before gunpowder was introduced in the seventeenth century. Later, after it was patented by Alfred Nobel in 1867, the safer dynamite became available. The larger the pillars, the more coal was left behind, whereas the smaller they were, the greater the risk of roof fall. This method persists in many mines worldwide, but in the UK had largely disappeared by the mid-twentieth century in coal mines, to be replaced by longwall mining. This is the type described above, whereby roads are driven, one each to either side of the face, the coal between being extracted by machine along the face and removed on a conveyor belt. This also allowed more efficient ventilation of the operation, air containing the dust and gases being extracted continuously from the face. The roof above the extracted coal is allowed to fall, and that above the miners is held up by wooden props, later replaced by movable hydraulic roof supports.

The increasing demand for coal through the seventeenth century compelled many mines to go deeper. The first deep mine recorded was that owned by Sir George Bruce in 1600, at Culross in Fife, up-river from where the Forth bridges now stand. He had two tunnels driven, one on an artificial island in the river and one on the shore, and his mine extended a mile under the water. Flood water drained into a well that was emptied by a continuous chain of buckets driven by three horses, very like the pumps illustrated by Agricola. Other such pumps were driven by water wheels; a beautiful example can still be seen at Laxey in the Isle of Man (Fig. 2.3).

Figure 2.3 The great 19th-century water wheel used for draining the lead mine at Laxey, Isle of Man. Hydraulic pressure fed the water by iron pipe to the top of the wheel, which operated a powerful pump.

Figure 2.4 Agricola's illustration of a trolley with iron guide pin (**F**) underneath.

As the industry expanded, the need for power to run the pumps and to transport materials increased. At some point in the seventeenth century, the railway was invented specifically for the coal industry, enabling trucks, drawn by people or horses, to run on rails conveying the product from the mines. As illustrated in Agricola's book, all previous transportation was by trucks and barrows using manpower (in coal mines, often womanpower) or horses. However, his trucks had an iron pin on the underside that ran in a groove in wooden planks to ensure they went where needed, presumably a forerunner of a rail (Fig. 2.4). The first record of rails in coal mining was in a colliery at Ravensworth in northeast England in 1671. These rails were first made of timber and were employed until iron became sufficiently cheap.

Early mechanisation: the steam engine

In 1698, Thomas Savery (*c.*1650–1715), an English military engineer, patented a steam pump which he promoted for raising water from mines – the first steam engine. The principle was a simple oval container which held water from the mine. Steam from a boiler was fed into this, driving the water out through a pipe. Then, opening and closing appropriate valves, the steam was condensed by cold water pouring over the cylinder, creating a vacuum that sucked more mine water in, and so on. Unfortunately these engines were inefficient, insufficiently powerful and because of a tendency to explode, too dangerous for the work required in deep mines. However, in 1712 the ironmonger Thomas Newcomen (1664–1729) patented a more efficient steam engine with a piston, based on an earlier model invented by a French scientist, Denis Papin (1647–1713), who had worked with Robert Boyle (see below) in London.

Newcomen's engine, first developed in tin mines in Cornwall, comprised a large open-topped cylinder with a piston that operated a beam attached at the other end to pump rods going down the pit. Steam was introduced into the cylinder to move the piston up, then condensed with cold water to bring it down again, the pump rods acting like a bicycle pump in pushing the mine water upwards. This proved more effective for driving pumps, being more powerful, having a shorter cycle time, and not requiring the boiler to be underground; it was first introduced into an English coal mine in Staffordshire in the same year it was patented, 1712. Its use rapidly spread through the British coalfields, and following improvements it proved sufficiently powerful to allow almost continuous extraction of water from the deeper mines.

In the early eighteenth century the iron trade had become depressed from a shortage of wood used to make the charcoal for smelting. Its revival was based on the use of coal or coke, first in Coalbrookdale in Shropshire, scene of the famous painting by Philip de Loutherbourg epitomising the industrial revolution. Improved technology of the furnaces, using Newcomen's engines, in this case to increase the supply of water for the blast furnace water-wheel, and the ready supply of cheap iron and coal led to rapid expansion of the iron industry throughout Britain (see chapter 4). From the mid-eighteenth century, cast iron steam engines with their cylinders and tanks became a prominent feature of British collieries. Power was needed not only for transport and for pumping water, but also to raise the coal from pit bottom to the surface. The first mechanical methods had involved using water wheels, sometimes with steam engines to pump the water to the wheel, until Watt developed a crank that converted the pumping motion into continuous rotation suitable for haulage.

Transportation: the age of steam

For transport, rails (from the 1770s, made of iron) were introduced both on the surface and underground, and horses rather than people were increasingly employed to pull the trucks. In the first years of the nineteenth century, steam engines based on Watt's improved designs replaced the less efficient and relatively dangerous Newcomen engines, which were prone to explode. They were developed further by a Cornish mining engineer, Richard Trevithick (1771–1823), using pressurised steam, to build the first steam locomotive. Initial problems related to the weight of the engines breaking the rails and the relatively poor traction obtained, and various methods were introduced to solve these, including rack and pinion railways and the use of chain traction.

A colliery manager called William Hedley (1779–1843) was the first to develop an engine that provided sufficient traction from iron wheels on iron tracks alone, hauling coal at Wylam near Newcastle-upon-Tyne in Northumberland. This engine became known as the Puffing Billy. Further improvements were made by George Stephenson (1781–1848), a largely self-educated engineer who had himself been born in Wylam. He built a steam engine called Blücher to haul coal in 1814 and with his son Robert, after producing a number of other colliery engines, went on to design and produce the famous steam engines *Locomotion* and *Rocket* in the 1820s. The age of steam, driven by coal, had arrived. It provided the power not only for mining machines but also for factories, trains and ships (the first of which operated as tug boats on the river Clyde in Glasgow), moving from small British engineering works across the Empire and the world. But the introduction of steam-powered machines into the collieries added another hazard to those faced by the workers – accidental injury from moving machinery. All advances in mining technology aimed at increasing productivity tended to bring with them new hazards for the workers.

Getting underground

By the mid-nineteenth century the increasing demand for coal had led to the need for deeper mines and better methods for raising the coal to the surface. Until this time wicker baskets called corves hauled by hessian ropes had been used. Miners, including children, descended in these or, worse, by holding onto the ropes, their feet in loops, and fatal falls were commonplace (Fig. 2.5). Thomas Hall from Ryton, a colliery manager who had started life as a pit boy himself, worked to

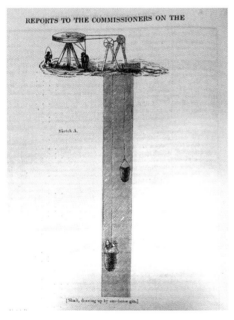

Figure 2.5 Illustration from Report of the Royal Commission of 1842 showing children being winched in baskets down mine shaft, using horse power.

develop a better method, first using iron tubs and later, in 1834, cages into which the coal tubs could be inserted, and in which the miners could be carried. At the same time, more robust iron wire ropes, which had been invented for use in the metal mines in the German Hartz Mountains, were introduced to haul the cages in place of hessian ropes.

Mine gases

At the time Agricola was writing his book in the sixteenth century, all that was known about air was that it was necessary to sustain life. It was believed to contain a vital principle that, when breathed in, was carried to the brain and nourished our immortal souls. The world was a mysterious place, as were the workings of our bodies. But miners knew a bit more about air than did the common man. They recognised that in some places underground you could suddenly fall unconscious or even die. Initially such events were attributed to demons or goblins, but eventually it became apparent that they were related to bad air, often though not always associated with fire. They spoke of 'choke damp' in stagnant air, the word being derived from the German *dampf* meaning gas or vapour. This led to a realisation that improved ventilation reduced their risks. In addition, explosions were a common occurrence, usually caused by the lighted candles that miners carried, of necessity. If they were lucky, a small explosion occurred, followed by fire running in the pit above their heads, but from the sixteenth century, as mines became deeper and more miners were employed, large explosions killing many workers started to occur more frequently. The air that ignited was called 'fire damp'. A third gas recognised was called 'stink damp' on account of its smell of rotten eggs; it was also known to cause asphyxiation.

The first scientific understanding of the air came from the experiments of Robert Boyle (1627–91), an aristocratic Irish natural philosopher who described the law of the inverse relationship between pressure and volume in gases, a phenomenon now well known to scuba divers; the deeper you dive, the more compressed the air in your lungs becomes: double the pressure, halve the volume. He showed in his vacuum jar that life of animals ceased when the air was evacuated. His apparatus was later used to show that only a part of the air, about 21%, was essential for life. The first individual gas to be discovered came from experiments by Joseph Black in Glasgow when investigating various chemicals as cures for indigestion. In 1754 he showed that on heating chalk (calcium carbonate) or treating it with acid, it lost weight and gave off a gas which he called 'fixed air'. That gas was heavier than air, did not support life, and extinguished a candle flame. Further experiments showed that the reverse could happen if he bubbled the gas through a solution of lime (calcium hydroxide). Although its chemical composition was not elucidated until the

English scientist John Dalton proposed his atomic theory in 1805, Black had discovered carbon dioxide. He set his student, Daniel Rutherford (1749–1819), later to become professor of botany at Edinburgh University, the task of discovering the properties of air after the combustible part of it had been removed; Rutherford called it noxious air, but we now know it as nitrogen, which comprises almost 80% of the air.

Three years later, in 1775, the English scientist and dissenting cleric, Joseph Priestley (1733–1804), published his discoveries of the properties of the life-giving substance in air, showing that it was given off by plants and was a better supporter of combustion than air itself. The Swedish pharmacist Carl Scheele (1742–86) had made the same discovery the previous year but published it later. It remained for the French scientist Antoine Lavoisier (1743–94) and his wife Anne-Marie to name it oxygen, and to show how animals derived their energy from it, converting it into the waste gas carbon dioxide in the process. This is the first half of what became known as the carbon cycle, to be completed more than a century later when it was shown by the German scientist, Justus von Liebig (1803–73), that plants did the opposite, converting carbon dioxide into oxygen. Priestley was also able to show that while combustion of oxygen produced carbon dioxide, in certain conditions of low oxygen availability another very poisonous gas was formed. Later this was shown to be carbon monoxide. As a further illustration of the international nature of eighteenth-century science, in 1776 the Italian scientist and inventor of the electric battery, Alessandro Volta (1745–1827), published his observations on a gas that bubbled out of Lake Maggiore and that did not support life, but was inflammable; he had discovered marsh gas, later called methane. A similarly toxic and inflammable gas with a smell of rotten eggs, the miners' stink damp (and the schoolchild's stink bomb), hydrogen sulphide, was another discovery made by Scheele in 1777.

Suffocation and explosion

The workers in eighteenth- and nineteenth-century mines did not of course know of these advances in chemistry. What they did know was that the air of mines could be dangerous in different ways; it could be suffocating or it could catch fire and cause explosions. Suffocation seemed to be a risk in stagnant air or in areas where fire had occurred, and this air sometimes lurked above the good air, so men climbing a ladder to a higher level might suddenly, without prior symptoms, fall unconscious, or one working in a poorly ventilated part of the pit might find himself becoming breathless. The major contribution to understanding these gases in mines was made by the Oxford professor of physiology, John Scott Haldane (1860–1936), one of the most famous scientists of his generation. His reputation came from his studies of the body's way of using and transporting oxygen. He also studied the effects on man of high altitude and diving, and in the First World War he had designed the military gas mask. We now know that fire damp is largely methane, a combination of hydrogen and carbon, usually generated by rotting organic matter and commonly found in coal deposits, where it can sometimes be heard hissing out of the seam; it is the commonest cause of fires and explosions underground.

Choke damp is either de-oxygenated air comprising nitrogen and carbon dioxide, somewhat heavier than air, or carbon monoxide, lighter than air and formed by incomplete combustion in reduced oxygen concentrations. Neither gas gives off an odour, so there is no warning, and both cause loss of consciousness and death if the victim is not evacuated and given oxygen rapidly. Such tragic episodes of death still occur in industry today, when workers inadvertently enter a tank or other space where such asphyxiant gases have accumulated; often such episodes involve multiple deaths as colleagues follow to rescue them and fall unconscious themselves. Stink damp, hydrogen sulphide, is also caused by rotting organic matter; in low concentrations the smell gives warning, but high concentrations may paralyse the sense of smell and cause asphyxiation. It is also inflammable and may cause explosions underground. To add to these risks, the terrifying propagation of fire in mines is enabled by the presence of coal dust, which in sufficient concentration can ignite and explode. Thus a small fire ignited by a candle or spark can send a fire sweeping along an underground roadway, causing multiple deaths.

Miners became aware of these dangers and paid attention to the flame of their candles, which might change colour or go out. Animals were used to go into poorly ventilated areas first, to test the air, and in some cases men called 'firemen' would go in cautiously ahead of their colleagues with a candle on a stick (probably the origin of the stick carried today by overmen) to test the air, wrapped in wet clothes in case of a fire or explosion. The last animals to be used in coal mines for this purpose were canaries, introduced by Haldane in 1911, as they would fall over at dangerous concentrations before people did, but then could revive quickly; they were only dispensed with in British mines in 1986, when chemical gas detectors became readily available.

Mine ventilation

Prior to the invention of engines, various methods were used to ventilate mines in order to remove poisonous gases and to ensure enough fresh air for breathing. Additional shafts were built and shutters or curtains were used to direct the flow of air towards where miners worked. Agricola described several different methods of ventilation in the sixteenth century (Fig. 2.6). These varied from devices to capture the wind above-ground and direct it through wooden ducts underground,

Figure 2.6 Agricola's illustration of a bellows method of ventilating mines in the 16th century.

to mechanical devices such as bellows and fans operated manually or by horses, to fires being lit at the bottom of the out-shaft to draw air down the in-shaft and round the workings.

As pits got deeper and more extensive, these methods became less effective, and more frequent explosions and suffocations occurred, with multiple deaths. Early in the eighteenth century it was noted that what were termed 'blowers' occurred in some mines, giving off inflammable gas, and methods were devised to shut them off. Much later in some mines these gases (if inflammable) were trapped and used as a fuel for surface machinery or for lighting. By the middle of the eighteenth century, as steam engines were introduced, steam-driven fans provided a more powerful source of ventilation and were used in association with what was called 'coursing' of the air to direct it by curtains and shutters to areas where men were working. Generally children were used to operate the shutters, sitting all day in pitch dark.

Aside from the need to reduce the concentrations of noxious gases in the mine air, ventilation also served to reduce levels of dust by dilution. The importance of this, explained above, became apparent in the nineteenth century when it was noted that explosions could ignite a fire that swept through the pit from the site of ignition, and the dangers of naked lights became very obvious. By the end of that century it had been established that ignition of coal dust was the cause of these devastating fires. Moreover, it had been shown, again by Haldane, that dilution of the coal dust by inert stone dust reduced the risks of these events, and the procedure of dusting the mines with limestone (which is not toxic to the lungs) was started in the early twentieth century. The risk of fire was shown to relate to the amount of inflammable matter in the coal dust, some coals being more dangerous in this respect than others. Coal was ranked in terms of the amount of heat it could produce, a high ranking indicating high combustibility, so mines that produced such coals, as in South Wales, entailed greater risks of explosion. These coal ranks became relevant also to the risks of lung disease, as discussed in chapters 7 and 8.

Deaths in mining and the miner's safety lamp

It is difficult nowadays to imagine a world without electricity, when artificial light was produced by candles and lamps burning vegetable or whale oil, and when a miner venturing deep into the earth had to use a naked flame in places where he knew there to be inflammable gases. By the start of the nineteenth century so many dreadful explosions had occurred, and so many lives had been lost that it became an urgent necessity to find a safe method of illumination. In particular, two explosions at mines in the northeast of England in 1812, taking a total of 126 lives, stimulated three people independently to work towards a safety oil lamp. A race began to produce a safe substitute for the naked flame, the principle being to allow sufficient air for combustion while preventing flame from propagating. The first investigator was an Irish physician, Dr William Clanny (1776–1850), whose original design proved impracticable. Next, the challenge was taken up by George Stephenson, the local man who was to design the *Rocket*, then an engineer in Killingworth colliery in the northeast of England, and who had personal experience of involvement in explosions. Simultaneously, the distinguished chemist Sir Humphry Davy (1778–1829), who became President of the Royal Society, also responded. Both men produced workable designs, safer though not very effective for lighting.

From about 1816 Stephenson's design, known as the Geordie lamp, was used in the northeast region of England and Davy's lamp elsewhere. There was dispute between them as to priority and the possibility of Stephenson having copied some aspects of Davy's design, but ultimately various improved hybrid versions were generally accepted. The principle that Davy introduced was to surround the flame by metal gauze, ultimately in a glass cylinder, which

Figure 2.7 A modern version of the miner's Davy lamp.

dissipated the heat from any ignition of gas that occurred within the lamp, thus preventing the flame from propagating outside and causing an explosion (Fig. 2.7). While the risk of explosion was reduced, none of these designs was completely safe and explosions still occurred. Two further inventions later paved the way for adequate illumination. In 1881 Sir Joseph Swan from Newcastle patented the incandescent light bulb, and in 1913 Thomas Edison in the United States invented a lightweight battery. Development of these inventions ultimately led to the modern helmet lamp powered by a pack on the miner's belt.

Moving into the modern era

From its earliest years mining was recognised as the most dangerous of occupations. The need for higher productivity through the nineteenth century had led to an increasing toll of deaths and serious injury among miners. In the UK, mines had to go deeper and coal seams became more difficult to reach and exploit. In 1850, the year in which a Mines Act was passed to introduce inspectors to mines, the beam of a pumping engine broke off and fell down the shaft of Hartley colliery in Northumberland, entombing 204 miners. This was the accident that led to an Act of Parliament in 1862 obliging all new mines to have at least two shafts.

Although major explosions and fires with multiple deaths attracted most attention in the press, there was a continuing weekly toll of individual accidental deaths throughout the year across the coal fields, each one a personal tragedy to the man and his family. Often these were caused by rock falls or transport injuries. Accidental deaths underground had occurred from 1900 at a rate of 1.2 per thousand employees *per annum* but had fallen to 0.86 per thousand at nationalisation in 1946 and 0.4 per thousand in 1960. However, even in its final decade in Britain, one man on average was killed every year, often due to accidents with moving machinery.

During the twentieth century, underground coal mining had become progressively more mechanised, mostly powered by compressed air and electric batteries, although some transportation was diesel. The mechanical coal cutter was first introduced in Lanarkshire in south central Scotland in 1907, but did not spread generally around the coal fields until the 1930s. In 1900 the 3384 British mines employed 780,000 men. In 1920, 2851 pits employed 1.25 million. This fell progressively thereafter to 958 pits at nationalisation with 704,000 men, reaching 65 pits with 57,000 men in 1980. After nationalisation of the British industry in 1946, mechanisation was increasingly introduced in a bid to maintain productivity. At the same time a national medical and first aid service was introduced with safety training for all miners. Conveyor belts, powered hand drills, rock-boring and coal-cutting machines, mechanical roof supports, and tunnelling machines were all introduced to replace the pick and shovel where the mine's geology allowed. Ultimately, continuous mining machines that cut the coal and loaded it in a constant cycle, and powered drilling machines to drive the roadways to the face were introduced where conditions allowed, significantly reducing the hard manual work of the miners and increasing the productivity of the pit. These more efficient methods of obtaining coal inevitably increased dust concentrations in the air of the mines and were probably responsible for the reawakened awareness of lung diseases in miners that started in the 1930s.

In particular, from the 1940s the danger of dust inhalation to the health of miners was increasingly recognised, particularly by the miners themselves, who had endured the higher dust levels that accompanied increased productivity. As control of the traditional dangers was slowly reduced, so the attention of mining engineers shifted to reducing the risks associated with breathing this dust, while that of doctors moved to understanding and quantifying those risks.

Chapter 3

The environment, disease and social reform

The dominant health issue in the early years of mining was the danger to life and limb from roof falls, floods, fires and explosions. These illustrated in a dramatic way the influence of one's environment on health; a miner's environment ensured that his or indeed her risk of early death was greater than that in any trade other than seafaring. But in the eighteenth and nineteenth centuries these risks were additional to the other risks that were shared with poor and disadvantaged people, and this led to life often being short and brutal. The early history of coal mining is that of an entrepreneurial industry with many small companies owned by wealthy landowners within whose property coal had been discovered. Competition was fierce and companies failed or amalgamated to form the larger concerns that could attract investment to finance the introduction of the new technologies as they were developed.

In the *laissez faire* economy of Britain in the eighteenth and nineteenth centuries,[1] exploitation of labour was the norm and family poverty forced children into work at a very young age. The pattern of work was changing, shifting from agriculture and small personal enterprises, such as the traditional crafts, handloom weaving, wood and metal working, to factory-based work, initially mechanised weaving. Changes in land use, particularly towards sheep rearing, forced increasing numbers of families from the country into the growing cities of Britain, especially Glasgow, Manchester, Birmingham, Leeds, Sheffield, Liverpool and London. Among them were my own ancestors, traditional handloom weavers in rural Perthshire, who moved to work in Glasgow. But in areas where there was coal, the mines often provided the only place of employment.

In order to understand the effects on the health of workers of the early history of coal mining, it is helpful to know something of the development in Britain of the concept of public health, as opposed to individual health, and of the influences of overcrowding and epidemics of infectious disease in the Industrial Revolution. The study of epidemics, epidemiology, became of fundamental importance to understanding the effects of work and, ultimately, of air pollution and climate change on people's health.

Overcrowding and disease

Between 1750 and 1850 the British population increased from around 6 million to about 18 million. The cleric and mathematician Thomas Malthus (1766–1834), in his *Essay on the Principle of Population* published in 1798, noted that population increased more rapidly than did food supply, and that this could thus lead to starvation and conflict. Such consequences can largely be avoided by increases in food productivity and control of family sizes, but to do so requires appropriate government action. Although Malthus has been criticised for his proposed solutions, this basic message holds true to this day – absence of good government combined with climate change and crop failure in Sudan and Rwanda remind us of this route to migration, warfare and genocide. Closer to home, in the 1840s the consequences of inappropriate *laissez faire* government were seen in the mass starvation and emigration affecting Ireland, then part of the UK, after the destruction by the parasitic alga, *Phytophthera infestans*, of the potato crop on which the poor had become dependent.

In the nineteenth century, housing accommodation in the towns and cities became hopelessly insufficient and sewage disposal was largely absent. Multiple families crowded into single buildings and even lived in one room. Household waste, urine and faeces were disposed of into the street, in some places to be collected by scavengers who transported it by horse and cart into the country to be sold to farmers as fertiliser.[2] The term 'slum', of unknown derivation, came to be applied to these areas of human overcrowding and poverty. In order to get an idea of what conditions were like in Industrial Revolution Britain, today it is necessary to visit some of the poorer world's megacities such as Calcutta, Rio de Janeiro, or Cairo, where gross poverty exists side by side with wealth, where population has outstripped the available sanitation, and governments are unwilling to redistribute the wealth necessary to provide it.

It is not surprising to modern eyes that infectious disease was rife in these circumstances and that mortality rates were high. Average life expectancy at birth was less than 20 years, owing especially to very high childhood mortality. The main killers were described in early mortality statistics mainly as symptoms, such as diarrhoea, fever, pox, and consumption or wasting. Some deaths came in epidemics while others were characterised by slow decline, but their causes were usually unknown and subject to speculation. Some were known to be passed from person to person in epidemics, but sporadic illness was commonly attributed to bad air[3] or miasma; in the absence of any useful cures there was a marked tendency for doctors and politicians to attribute disease to personal habits which were recognised to be harmful, particularly alcohol or an immoral lifestyle.

The naming of diseases

Gradually, following the 1665 Great Plague of London, the understanding of diseases increased and they acquired specific names through the eighteenth and nineteenth centuries. A London doctor, Thomas Sydenham (1624–89) led the way, clarifying the distinctions between epidemics of measles, smallpox and scarlet fever.[4] A highly fatal febrile disease in which the victim lapsed into coma became known as typhoid fever from *typhos*, a Greek word meaning smoke or stupor.[5] Smallpox, another often fatal febrile disease characterised by multiple pus-filled skin lesions, was differentiated from the great pox or syphilis. Syphilis was known to be sexually transmitted and was characterised by an initial genital lesion followed later by skin eruptions. It was found to lead to a fatal brain disease (general paralysis of the insane, GPI); even up to the early twentieth century this was responsible for most cases of insanity in mental hospitals. It could also lead to rupture of the aorta, the main blood vessel from the heart. Women commonly died in childbirth of puerperal fever even after it was recognised, in Aberdeen and later in Vienna, that it could be prevented by attendants washing their hands.[6] The often fatal diphtheria in children was easily recognised by an inflammatory membrane in the throat and high fever. Consumption, also called phthisis, which described a wasting condition associated with cough and bloody sputum, was named tuberculosis after the characteristic pathological features, tubercles, were described in the lungs in the nineteenth century.

Cholera, a usually fatal diarrhoeal disease, spread from India around the world in the nineteenth century, arriving in Britain in the first epidemic in 1831. Its name, meaning bile, harked back to the time of the ancient Greeks and Romans when most diseases were attributed to imbalance of the four 'humours': black and yellow bile, blood and phlegm (from which we get the terms melancholic, choleric, sanguine and phlegmatic). Before we laugh at the naïvety of such beliefs we should recall a similar tendency in the early twenty-first century for some enthusiasts to attribute all disease to our genes. Both concepts are attempts to explain the general background on which our environment acts to make us who we are and become, in the light of current scientific understanding.

These epidemic diseases, together with the more familiar and less frequently fatal childhood febrile diseases, measles, scarlet fever and mumps, became the spectre that stalked the people living in Industrial Revolution Britain. All proved later to be the result of infection, but bacteria and viruses were then unknown. Interestingly, because mining communities were initially based mostly in villages in the country with less overcrowding, the background risk of these infectious diseases was somewhat less severe than in the mills and workhouses of the cities. Childhood mortality was lower and country folk on average lived longer lives than did town dwellers.

Understanding the causes of disease and their prevention

It has to be admitted that the medical profession was of very little use to the poor until the middle of the nineteenth century. Not only were doctors generally interested only in those who could afford their fees, but also most treatments offered were either ineffective or, worse, positively dangerous. As blood-letting, purgation and induction of vomiting were the physician's mainstays, it could be argued that the poor were fortunate to be spared medical attention.

A few notable exceptions to this general rule did occur, however. One of the first doctors to note effects of work on health was Agricola, mentioned in the previous chapter, who noted the poor conditions of miners in the sixteenth century leading to premature mortality. He recorded that some women in the metal mining area in which he worked had been repeatedly widowed and that consumption was rife among miners. In his day the term 'consumption' applied to all wasting diseases, but two miners' diseases then unknown, silicosis and lung cancer, were also likely to have contributed. The first major step in understanding the importance of the workplace in causing disease was taken by Bernardino Ramazzini, (1633–1714) professor of medicine first in Modena then in Padua, Italy, when in 1700 he published his book *De morbis artificum diatriba* (Thesis on the diseases of workers). His interest was aroused by a conversation with a labourer cleaning out the cess pit of his house, and he was probably the first teacher of medicine to draw doctors' attention to the importance of asking patients about their work. This message was taken up over 100 years later by a doctor in Leeds, Charles Turner Thackrah (1795–1833) in an analogous book published in 1832, *The effects of the principle arts, trades and professions on health and longevity*. By this time it was becoming apparent that the conditions in which the poor lived were likely to be influential in determining their high mortality and, while doctors in general had a tendency to blame their patients for their illness, as clergy might attribute misfortune to sin, some reformers began to investigate housing and social conditions.

Medicine was groping around in the dark in the early nineteenth century. Treatments were mostly symptomatic, although a few useful drugs were available – digitalis derived from foxgloves for heart failure, opium from poppies for pain, and quinine from South American tree bark for fevers and malaria. Until the causes of the killing diseases could be found, there was little hope of prevention but, one by one, small but important breakthroughs began to appear. An early indication of the concept of infection had come from recognising contagion, contact, in spreading smallpox; as early as 1717 Lady Mary Wortley Montague, wife of the British consul in Constantinople, had recorded the use of inoculation in Turkey to prevent smallpox. She had her own daughter inoculated, and this became popular among the British aristocracy. The crude initial method of injecting pus from smallpox victims into children's skin was subsequently improved and put on a scientific footing by Edward Jenner (1749–1823). He surmised that a similar but much milder disease of cattle that was occasionally transmitted to milkmaids, cowpox, might prove an effective and safer alternative source of pus. His book published in 1798, *An inquiry into the causes and effects of the variolae vaccinae,*[7] attracted worldwide attention and even Napoleon had his army vaccinated. The first British law to prevent disease, requiring vaccination of babies, was passed in 1853, although it was not enforced.

Miasma or germs? The beginning of epidemiology

Smallpox was but one of many diseases that occurred in epidemics and were thought by some pioneers to be contagious; at that time such epidemics were responsible for some 20,000 acute deaths each year in the UK. A second major advance in understanding came in 1855 when the physician John Snow (1813–58) in London completed his study of the 1854 cholera epidemic. By coincidence, my great grandfather, who had broken with the family tradition of handloom weaving, a trade then being displaced by the arrival of factories, was involved that same year as a medical student in treating cholera patients in the Glasgow area. Contrary to the contemporary belief that the disease was airborne, a condition caused by miasmas, Snow thought it might be carried by water, based on the fact that it affected the intestines rather than the lungs. He had been born in poor circumstances in York but apprenticed to a surgeon in Newcastle at the age of 14, working in the mining village of Killingworth, where he was involved in looking after miners and their families with cholera during the 1831 epidemic. He continued his studies in London and published a proposal that drinking water rather than miasma was responsible. In a landmark study of cases of cholera occurring in the region of Broad Street in the Soho district of London, he demonstrated that far more cases of the disease occurred among people who drew their water from the local pump in Broad Street than did among those who obtained their water from elsewhere, notably in a local prison which had its own well. As the local pump was owned by a company that drew its water from the heavily contaminated Thames, these observations supported his theory, and he reported it to Parliament, urging that there should be significant improvements in drainage and sewage disposal.

Within a year of Snow's report, a physician in Bristol, William Budd (1811–80), drew a similar conclusion in relation to the transmission of typhoid fever. In the absence of knowledge of the existence of bacteria, medical thinking began to consolidate around the presence in faecal matter of so-called zymotic particles sustained by decaying organic matter.[8] Budd's book on typhoid fever details the studies of outbreaks that he carried out, stimulated by personal experience of treating patients and seeing families dying and leaving poor orphans. He was able to show that disinfection (a term he used before the germ theory of disease) of excreta was able to cut short outbreaks, but his account makes clear that his views were strongly opposed by senior members of his profession, who still adhered to the miasma theory.

Coincidentally, by the 1830s fungal organisms had been found in plant diseases and even in a skin disease, ringworm, and the concept of minute living organisms being responsible for disease gained some credibility. This was in spite of continuing opposition by the conservative medical profession, who even managed to delay progress in preventing cholera and typhoid. However, the incorrect theory had one beneficial outcome, for it was the offensive smell of the polluted Thames that became too much for members of the British Parliament. The engineer Joseph Bazalgette (1819–1890) was commissioned to build a sewage and drainage system for London, drinking water being obtained from the upper Thames and passed through filter beds. This was completed in 1875, marking the start of the sanitary revolution in Britain. By this time, Louis Pasteur (1822–1896) in Paris had demonstrated the presence of microorganisms and their roles in fermentation and putrefaction; in 1878 he presented his germ theory of disease to the French Academy of Medicine.

These accounts of the investigations of populations, rather than solely individuals, documenting their health episodes and associating these with potential causes, represent the start of scientific epidemiology. The hypothesis that water carried the causative agents of cholera and typhoid was supported by the results and led to the conclusion that the diseases could be prevented by practicable measures. The truth of this conclusion would eventually be confirmed by both demonstration of the causative organisms, which could then be investigated in laboratory experiments, and most importantly in actual prevention of the diseases in populations by appropriate action. Application of this logic was to play a major part in understanding the various effects of coal on miners' health.

Preventing disease by legislation and regulation

It would be wrong to characterise the entire medical profession in the eighteenth and early nineteenth centuries as insensitive to the conditions of the poor. At that time the indigent were subject to the Poor Laws, which provided some support and accommodation in workhouses; these became overcrowded and sites of disease transmission and death. In Manchester and Liverpool volunteers set up local health boards. The first attempt at regulation was the *Health and Morals of Apprentices Act* of 1805, introduced by Sir Robert Peel (father of the Prime Minister of that name), after he discovered the conditions of children working in his own factories in Lancashire. Peel's Act was urged on by Dr Thomas Percival

(1740–1804) and the newly formed Manchester Board of Health. Percival had been orphaned in early childhood and brought up by an older sister in Lancashire. He attended medical school in Edinburgh and practised in Manchester, where he was a founding member of the famous Literary and Philosophical Society. He was the first to write a system of medical ethics and was much influenced by the views of a fellow member, Robert Owen, who was shortly to introduce an eight-hour day for children in his cotton mill at New Lanark. The 1805 Act obliged employers to introduce some reading and arithmetic lessons with instruction on Christianity, and also restricted hours of work to 12 per day. However, it introduced no enforcement and only applied to those children taken on as apprentices, ignoring those orphans who were effectively enslaved; it was thus a dead letter. Laws and regulations require oversight by people with the power to enforce them, an obvious fact that is still sometimes ignored by politicians.

A leading light of the period was the philosopher Jeremy Bentham (1748–1832), the founder of Utilitarianism, which sought the greatest happiness of the greatest number, a concept derived from the Professor of Moral Philosophy in Glasgow, Francis Hutcheson (1694–1746) and the early years of the Scottish enlightenment. To achieve this objective, a mathematical approach to the study of populations was necessary. Bentham took the lawyer Edwin Chadwick (1800–90) as his assistant, influencing him on his path to becoming a leading social reformer. Chadwick became secretary to the Commission that had been set up to review the working of the Poor Law.

Realising that disease played a major contributing role in poverty and thus claims on the available support funds, Chadwick appointed doctors to investigate the situation of the poor in London, one of whom, James Kay-Shuttleworth (1807–1877), had already published a study of the conditions of the Lancashire cotton workers. Their initial report pulled no punches, leading Chadwick to extend his survey to the whole of Britain. The *Report on the sanitary condition of the labouring population of Great Britain*, published in 1842, emphasised the influence of overcrowding and absent sanitation on health and poverty and led to the development of local boards of health and appointment of medical officers of health in the major cities, Liverpool being the first with William Duncan (1805–63).

Duncan set up the first city public health board with the borough engineer and an inspector of nuisances, demonstrating the effectiveness of cross-disciplinary team work in reducing illness in communities. The first *Public Health Act* of 1848 was enacted to oversee the reforms, including the regulation of offensive trades, sewage and water supply. In view of the role of alcohol in predisposition to disease, it is ironic that both John Snow in London and William Duncan in Liverpool are commemorated in the names of public houses.

Early mines acts

The situation of the mines differed from that of factories in the cities, since mines were sited in villages and country districts. Here there were two important issues: the exploitation of children and women and the repeated and increasing episodes of explosions and fires resulting in considerable loss of life among miners. The same issues affecting health and longevity that Chadwick's report had identified in the cities applied among the poor families in mining areas, although with less overcrowding in the country there was less transmission of what would become recognised as infectious diseases. As the coal industry expanded to satisfy the demands of other industries, mines became deeper and the numbers of workers required increased. Driving the pits through the water levels increased the risks of flooding and in some mines, recognised by the workers as gassy, the risks of explosions increased. It became apparent that the safety lamp was no guarantee of absolute safety, and indeed in high gas conditions could ignite an explosion. A flood in Heaton colliery in northeast England in 1815 took 75 lives and attracted public sympathy but no concrete action. Continuing explosions and deaths attracted the attention of Parliament, which set up a Select Committee to look into means of prevention in 1835; ironically, within a few months a further huge explosion in Wallsend colliery killed 102 miners. In spite of this, the Committee felt unable to make any helpful recommendations in terms of regulation.

By the 1840s some 250,000 people, a quarter of them women and children, were employed in coal mining, many having transferred of necessity from work on the land. However, the dreadful conditions and the employment of women and very young children (Fig. 3.1) in this trade passed unremarked by legislators until a disaster occurred in which 26 girls and boys drowned in a mine in 1838. Shocked by this, the young Queen Victoria, who had ascended the

Figure 3.1 Woman and two children dragging wheeled coal tub to the foot of the shaft c.1780 (Google image: www.industrialrevolution.org.uk).

throne in 1837, ordered an inquiry led by Lord Anthony Astley Cooper, later 7th Earl of Shaftesbury. The consequent report of the *Royal Commission on Employment of Children in the Mines* was published in 1842. The illustrated verbatim stories of these women and children make distressing reading; they were in servitude to their masters – truly slaves. The report affected Parliament so much that regulation, the first Mines Act prohibiting employment of women and children under 10 years old, was introduced the same year – a small step in the right direction. It is said that the decision to illustrate the report with etchings was a decisive factor, as the depiction of half-naked women and men working together underground and of six-year-old infants being winched down shafts by their mothers stirred the consciences of the members of Parliament.

In his speech to Parliament, Lord Astley said:

> Surely it is evident that to remove, or even to mitigate, these sad evils will require the vigorous and immediate interposition of the legislature. That interposition is demanded by public reason, by public virtue, by the public honour, by the public character, and, I rejoice to add, by the public sympathy: for never, I believe, since the disclosure of the horrors of the African slave trade has there existed so universal a feeling on any one subject in this country, as that which now pervades the length and breadth of the land in abhorrence and disgust of this monstrous oppression.

Aside from the prohibition of young children and women from working underground, the Act also recommended a mines inspector; although he was not to go underground himself, he was to report to Parliament annually, leading to Government taking note of accidents subsequently. Explosions continued to occur throughout the 1840s, and each one that involved substantial loss of life resulted in an inspection by scientists and a formal report; most of these pointed to three issues: poor ventilation, careless use of lights, and the need for preventive underground inspections. It was also noted that although major losses of life attracted most attention, an even greater number of deaths occurred in small numbers almost continuously, and that conditions and safety measures differed considerably between mines.

In 1850 Parliament at last conceded this point and introduced an *Act for the Inspection of Coal Mines in Great Britain*. This act was subsequently strengthened every few years. In 1862, following the deaths of 204 miners in the Hartley colliery disaster, mentioned in the previous chapter, an act prohibited mines from working with a single shaft, so that all mines were required to have an additional safety shaft. With the comparable acts to protect children from exploitation, this marked the start of legislation that would very slowly erode the power of employers in UK to exploit workers, by introducing a mechanism for policing compliance.[9]

This remains a fundamental issue of politics to this day, and a balance in Britain was only struck when the coal industry was nationalised in 1946. Prior to nationalisation, an average of 1.1 miners were killed at work per 1000 employed each year, that is about 160–170 deaths annually, of which 70% were caused by roof falls and about 25% by haulage accidents. These figures were very similar to those in other European countries and, surprisingly, in British India, but matters were much worse in Canada (1.9 per 1000), USA (3.55 per 1000) and South Africa (3.5 per 1000).

While the late nineteenth and early twentieth centuries saw significant improvements in the conditions of coal miners underground, it should not be forgotten that their livelihoods were very much at the mercy of their employers and of fluctuations in the market for coal. Some mines owned by benevolent landowners provided a measure of protection for the men in hard times, whereas others owned by absentee landlords or by entrepreneurial companies that had bought the mineral rights provided none.[10] It was only when various early attempts to provide benefits for the unfortunate were consolidated into the Welfare State in the UK after the Second World war that workers were afforded better protection from such vicissitudes of fortune.

Compensation for injury and disease

From the above brief account it will be clear that coal mining continued to cause a huge amount of death and disability among the workforce throughout the nineteenth and early twentieth centuries, for almost all of which there was initially no system of compensation other than taking legal action against the employer. An *Employer's Liability Act* of 1880 made clear that the burden of proof lay with the injured worker, making sure that this course was unrealistic for the impoverished workers of the time. Finally in 1897 the first *Workmen's Compensation Act* was passed by the British parliament, applying initially to those workers employed in mines, quarries, factories, railways and laundries. It obliged the employer to pay the compensation according to a fixed rate based on earnings, and made clear that employers would be well advised to insure themselves against such costs.

In 1906 this act was replaced by one that extended the right to all employed workers injured in their job, save those in non-manual occupations earning more than £650 per annum and certain casual workers. However, by this time it was becoming apparent that many workers were at risk of disease as well as accidents at work, and a Committee on Compensation for Industrial Diseases was established under Herbert (later Lord) Samuel to decide what diseases might be distinguishable as a consequence of occupation. In this review it was decided that although lung disease, described as 'miner's phthisis', did occur its features did not allow it to be distinguished from phthisis acquired otherwise than at work, and it could not therefore be included. However, recognition of a similar condition in South African gold miners led to specific compensation being payable to such workers in 1912 in that country, then part of the British Empire, and after further investigations a similar *Workmen's Compensation (Silicosis) Act* was finally introduced in Britain in 1919. How far this applied to coal miners is discussed in chapter 6.

In summary

The nineteenth century in Britain was a period of stark contrasts, to which the story of coal was central. Coal enabled the Industrial Revolution and the revolution in transport of goods, and the ensuing prosperity enabled better nutrition through developments in agriculture. In turn, this allowed more children to survive the prevalent infectious diseases and to join the increasing workforce in the enlarging cities. At the same time, the overcrowding and the exploitation of labour, especially of children, awakened the consciences of some reformers, and more enlightened legislation began to take account of the disenfranchised and exploited poor. This was the start of what we now recognise as the responsibility of Government to the whole population, and led eventually to the Welfare State of the 1940s. The nineteenth century also witnessed the start of preventive medicine, with use of information from epidemiological studies to plan methods of protecting the population from disease and accidents in the overcrowded cities and in factories and mines. In the next chapter we look specifically at the role coal played in defining the character of the world we have been living in: how coal became king.

Chapter 4

King Coal: the rise to power

As children most of us take the things around us for granted, though a natural curiosity leads us to ask questions. Today's children accept the apparent miracles of electronic technology, robots and drones, as we accepted motor cars and aeroplanes, but at some stage in our lives curiosity leads us to try to understand them. In the late 1940s my father bought our first house. The coalman came every month or so in his truck. His face and arms were black and he wore a leather apron. The back of his truck contained hessian sacks of coal and he carried these to the side of the house to empty them into a shed. Some of the larger houses in the area had cellars to accommodate the coal. I began to wonder about the story behind this black material.

In 1950 we had moved to live in a town called Birkenhead, across the river Mersey from Liverpool. Birkenhead was known for having had the first public park and electric tramway in Britain and for pioneering the manufacture of iron ships. The park, designed by Joseph Paxton who went on to design the Crystal Palace, was built to give city dwellers a space where the air was relatively free from the black smoke from domestic and industrial chimneys. It had been visited by Frederick Law Olmsted on his visit to Britain from America in 1850, and he took it as inspiration for his design of Central Park in New York, in its turn the first public park in the USA. In these ways, Birkenhead had epitomised the importance of coal and steel in the rising Western economies of the nineteenth century.

In the 1950s, across the River Mersey in Liverpool trams still ran on electricity generated by coal-burning power stations. Like all houses at that time, we had electric light and piped gas for cooking. The street lights were electric although a few years before they had been lit by coal gas, and men had been employed to light them as night fell. Not far from our house there was a gas works that produced a lot of smoke and a distinctive smell. Ferries powered by coal took us across the River Mersey, or an electric train took us under it, to Liverpool where we got a noisy, rattling electric tram from the Pier Head into the centre of the city. Electricity was beginning to be used in the burgeoning music industry, where a new group with electric guitars was making a name in the ill-lit Cavern Club; clever name, I thought when I first saw one of the Beatles' hand-printed posters in the late 1950s.

In our house, several rooms had fireplaces with iron grates and there was an iron boiler in the kitchen. A chimney sweep visited periodically on his bicycle to clear the chimneys with the brushes that he carried on his shoulder. This was necessary, as the soot in the chimney could ignite and was a cause of houses burning down. The kitchen implements were all made of steel, the plates of pottery or china. A milkman delivered milk in glass bottles each morning. Most of the furniture was made of wood, including the casing of the wireless and the gramophone, but there were a few plastic materials in the house – a Bakelite telephone and gramophone records. All these things we took for granted, but behind them all was coal, even powering the kilns in which the glass and pottery were made, the looms that had woven the materials that we wore and that covered the furniture, and the new plastics. Not only did it provide us with the fuel for heating; it was also the source of both gas and electricity. It powered the industries that manufactured almost everything else and was the feedstock for many of the chemicals and even some medicines such as paracetamol that we had in the house. Coal, by far the dominant source of power, was king.

What is coal?

Most people of my age remember coal as lumps of hard, black and fragile stone, easily sheared by a blow from a poker. It required a fair bit of heat from kindling wood to ignite, but once lit would burn brightly and finish up as a pile of grey dusty ash. This was its traditional use. The heat energy came from its carbon content, derived from the trees and other plants that formed it. Often an oily substance would ooze out of it as it burned, and occasionally you could hear a hiss of gas issuing from a lump, to be ignited in the fire. Coal may contain all these things: carbon, oils, gas and ash. The ash comprises minerals called silicates, clay-like matter with names such as kaolinite and mica, together with crystalline silica (quartz) itself. There are also other chemicals in small amounts, such as sulphur, arsenic and mercury.

Coal differs in its detailed composition from pit to pit, from seam to seam and from country to country. It may be classified as hard coal (anthracite), steam (bituminous) coal, and soft coal (lignite), in descending order of combustibility. Conventionally, coals are assessed by rank according to their carbon content and thus combustibility. The composition of coal is important to industry, different coals being suitable for different applications. In addition, the composition is relevant to the pollution that coal causes when it is burned and to the lung diseases that miners get when they inhale its dust. If, after a miner's death, you were to obtain a piece of his lung at autopsy and then turn it to ash, you would find that the minerals in it, even many years after he had stopped work, would match those found in the coal seam in which he had laboured. These differences could even be reflected in the pattern of disease in his lung, as will be discussed in chapter 8.

Iron, steel and coke

Coal provided the essential power for the Industrial Revolution, but all the mechanical artefacts that enabled the rise of the great economic powerhouses in the western world were made of iron or steel. In chapter 1 we imagined a walk along the River Almond past the ruins of water mills where iron had been smelted and fashioned into farm tools and nails. At Cramond on the Firth of Forth, we looked west towards the famous Forth Bridge, constructed of steel in the 1880s and opened in 1890 to carry the railway from Edinburgh north to Dundee and Aberdeen. It was at that time the longest cantilevered structure in the world and was the tallest man-made structure, save for the contemporary Eiffel Tower. Should you have boarded a train crossing the bridge you would quickly reach Dunfermline, which you can see across the river, the birthplace of Andrew Carnegie (1835–1919). Carnegie, like his Scottish contemporary, my great-grandfather, was born the son of a handloom weaver. His father fell on hard times in the 1850s and emigrated with his family to Pennsylvania. The son Andrew worked as a telegrapher from the age of 13 and was self-educated, working his way up in the railroad industry and progressively amassing a fortune that led to his playing a major part in the story of coal with his iron and steel works in Pittsburgh.

The story of iron and steel is integral to the story of coal. Iron ore is found in many countries, the iron being combined with oxygen as iron oxides.[1] From about 800 BCE, the start of the British Iron Age, iron was smelted by crushing the ore and heating it with a source of carbon, usually charcoal, to combine the oxygen molecules in the iron ore with carbon, producing carbon dioxide as a waste gas. The metallic iron was separated from impurities by addition of lime, originally from sea shells, and cast into moulds; the blacksmith would then hammer it into shape to produce wrought iron. From the fifteenth century the process was improved by the development of the blast furnace, whereby air was blown through the molten mixture, the same principle that we used when as scouts or guides we blew on smouldering sticks to encourage them to flame. The metallic iron thus formed could then be poured into moulds as cast iron.

In 1709 Abraham Darby (1678–1717), a pioneer in casting brass and iron pots, had discovered that certain bituminous coals could be baked in a furnace to remove their sulphur and tar to produce coke, which was even more useful as a source of carbon in making iron. As we shall see, the tar also had its uses. As charcoal became scarce from overuse of trees in Britain towards the end of the seventeenth century, coke became a major product of the coal industry. But neither form of iron, wrought nor cast, was suitable for all applications. Steel, which is iron with a carbon content of around 1%, provided the answer, being intermediate between wrought and cast iron in carbon content and being both durable and flexible. The ability to control the amount of carbon in steel, and thus to design its properties, came in the second half of the nineteenth century with the inventions of the Bessemer converter and the open hearth process for making steel from iron. From then on coke poured into the steel furnaces and steel built the railways, the bridges, the trains, ships, vehicles and skyscrapers, as well as our knives and forks and weapons of war. And Andrew Carnegie in Pittsburgh made his fortune, having moved from a handloom weaver's cottage in Scotland to become the world's richest man.

Let there be light!

We take so much for granted. Most mornings I go into my dark study and switch on the light. If it is cold, I turn on the central heating before sitting down to work on my computer. And I do this most days without thinking of what lies behind all these things until the fuel bill comes at the end of the month. As a doctor I was aided in examining my patients by using several simple battery-powered instruments to examine eyes and ears, as well as by X-rays. Then came CT scans and multiple automated chemical analytical machines to replace the simple chemical tests that were once carried out in a side room of the hospital wards in order to investigate the patients further. Imagine what life would have been like before electricity and gas. Imagine a doctor, for example, examining a child on a dark winter's day using a candle for light, trying to look into her ear or down her throat, or a harassed housewife trying to darn her children's clothes by candlelight.

For most of its history coal was used solely as a source of heat, thermal energy, powering the factories of the Industrial Revolution and heating the large houses of the wealthier people. However, it provided a further impetus to industry in the late nineteenth century from the introduction of improved lighting using coal gas, allowing longer hours of work in the winter months. This was followed by a revolutionary change to the ordinary lives of almost everybody – the winter day became as long and productive as a summer's day. William Murdoch (1754–1839) is less well known than he should be. Like James Watt, he was born in Scotland and early in life he also showed that he had an inventive nature and an aptitude for mechanics. He became interested in steam engines and, aged 23, walked the 300 miles to Birmingham to ask Watt for a job in his and Matthew Boulton's steam engine company. He became their senior engine erector. In 1779 in this capacity he was sent down to Cornwall, where the company was heavily engaged in producing engines for the tin mines. He made significant improvements to the engines and was behind several of Watt's patents. He is also credited with having invented the first steam-powered prototype of a vehicle that could travel on roads, although he did not patent it.

In those times street lighting was absent, though various laws had been enacted to try to compel householders to hang lanterns outside their properties. While he was in Cornwall, Murdoch experimented with various methods of producing gas and found that by heating powdered coal he could produce a gas that, when ignited, gave a bright light; he saw the potential of using this for lighting, replacing the use of oils and tallow candles. By 1794 he had devised a system of heating coal to produce gas and distribute it through iron pipes, and his house in Redruth was the first to be lit by this means. Again, he missed the opportunity to patent the discovery. Later, on his return to Birmingham, he introduced gas lighting to the Boulton and Watt engine factory and to their Soho foundry in Smethwick near Birmingham in 1798. An employee of this foundry, Samuel Clegg, formed a company to exploit it. Gas lighting was demonstrated in Pall Mall in London and the first street with gas lighting was Westminster Bridge over the Thames by the Houses of Parliament in 1813. By 1816 street lighting had spread to Preston in Lancashire and to Boston USA, and by the mid-century was to be found in cities and towns round the western world. Only in the second half of the twentieth century was it displaced from the streets by electricity; again much of this was produced in power stations by burning coal until, in the 1970s, it started to be displaced by natural gas from the North Sea and further afield.

The development of a pipe network also allowed coal gas to be progressively adopted in houses for lighting, in London from the 1820s and by the 1840s in most cities and towns in Britain. Its use for cooking lagged two decades behind but received a boost when gas ovens were shown at the 1851 Great Exhibition in London. These advances, however, brought with them two hazards: explosions and accidental or suicidal poisonings. Indeed, until the much less poisonous natural gas (largely methane) was substituted for coal gas (20–30% carbon monoxide), putting your head in a gas oven was an important and readily available means of committing suicide.[2] Domestic natural gas displaced coal gas in the USA in the 1930s and in Britain from the 1970s after the discovery of the North Sea deposits. However, its use for domestic and street lighting had declined from the time an electricity grid was built in the 1930s, since electricity was much more easily introduced into new housing developments.

One residual use of coal gas remains. Syngas, short for synthesis gas, is produced by passing steam through incandescent coke, the main constituents being carbon monoxide and hydrogen. It contains the essential chemical elements, carbon, hydrogen and oxygen, for use as building blocks for larger molecules and is used for making hydrogen itself (with potential as a clean vehicle fuel), ammonia for fertilisers and a very wide range of chemical products, following the discoveries in organic chemistry as discussed below.

Coal and the origins of organic chemistry [3]

The story of chemistry and the chemical industries that have transformed our lives is intimately connected to man's use of fuel for energy. Early mankind unknowingly used chemical reactions to produce bronze, by using charcoal to remove oxygen from copper ore and then by combining the metallic copper with tin ore. Later, iron was produced in a similar way, ultimately using coal and coke as described above. Using heat to donate energy to chemical reactions is as old as mankind, and can be traced from the furnaces of the alchemists of the Middle Ages, attempting to transmute base metals into gold, to the Bunsen burners using coal gas that we first encountered in our school laboratories. But the use of fire in early experiments on dry minerals would have destroyed any organic matter, that is, substances based on carbon, which would have been oxidised by air into carbon dioxide. Organic chemistry, the chemistry of carbon compounds, owed its origins to coal itself.

The three major coal products were gas, tar and coke. Many coal mines had associated coke plants where the bituminous coal was baked in ovens at high temperature in the absence of oxygen to produce coke. These ovens were notoriously polluting, since the purpose of the process was to force out the volatile constituents of the coal. Both coke and gas production left a tarry residue, and from the eighteenth century uses started to be found for this, initially for waterproofing wood. Distillation of the tar produced a variety of different materials,[4] the chemicals benzene, toluene, naphthalene and anthracene, together with creosote and pitch.

Charles Macintosh in Glasgow in 1819 used naphtha to dissolve rubber and join two layers of woven material together to make what became called the mackintosh. In 1856 William Perkin produced the dye mauve, from aniline derived from the distillate, and initiated the synthetic dye industry that flourished for the next half century in England and France. These chemical colours eventually replaced the use of natural dyes derived from plants, molluscs and insects that had traditionally been used for colouring materials and as pigments for paints.

The production of dyes from coal products marked the start of organic chemistry, in contrast to non-carbon based inorganic chemistry. The latter derived from medieval alchemy, and was essentially based on manipulation of minerals with heat, acids, and alkalis to produce useful materials such as glass, soap and fertilisers. Substances containing carbon were initially thought to be derived only from living matter – hence the name 'organic'. Then, in the late nineteenth century, it

43

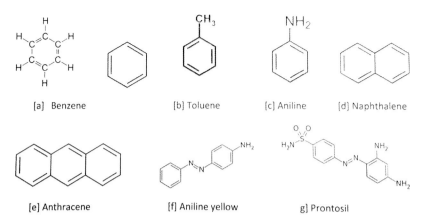

Figure 4.1 The origins of organic chemistry: benzene [**a**] is a ring of six carbon atoms, each linked to a hydrogen atom, usually shown as a hexagon. Chemicals [**b**–**e**] are also coal distalation products. [**f**] and [**g**] are formulae of the first organic dye and the second chemo-therapeutic substance respectively, showing how they are based on the analine molecule.

became possible to understand the structure of organic molecules and to make them in the laboratory from simple substances such as aniline, carbon monoxide, water, and ammonia (Fig. 4.1). It was at this point that the chemical industry in Germany took over from the English and French experimentalists. The German chemical industry was centred round Justus von Liebig (1803–73), best known to schoolchildren studying chemistry from the condenser named after him. He was a professor in Giessen and later in Munich. He had discovered that artificial fertilis-ers could be made by treating bones with acid, and had set up a school of chemistry to which aspiring chemists and entrepreneurs from all over Europe and even the USA came to learn from him.

Out of this school came an understanding of the biochemistry of plants, espe-cially their dependence on carbon dioxide for photosynthesis (thus their role in completing the carbon cycle mentioned in chapter 2) and their need for nitrogen, potassium and phosphorus for growth. Liebig had shown that artificial fertilisers could be obtained from coal gas production by combining the ammonia produced as a by-product with sulphuric acid to make ammonium sulphate.

Liebig's arrival in Germany after his own studies in France coincided with the rise of the gas light industry and thus an increasing availability of the residual tar. Distillation of this was used by the chemical industry to provide feedstock for the synthesis of dyes, and later of medicines. It was the origin of the great German chemical companies such as Bayer, Hoechst and BASF. From this came an under-standing of the aromatic organic chemicals based on a ring structure, the simplest of which is benzene with six carbon atoms, each coupled to an atom of hydro-gen, arranged in a ring (Fig. 4.1). Two rings coupled together make naphthalene and three together make anthracene. By 1850 it was known that all three of these chemicals could be extracted from coal tar, together with the simpler structures phenol and aniline. Later, benzene was also extracted commercially from the waste gases of coke plants in steel works. Understanding that such molecules (combinations of atoms) could behave like single chemicals and be combined with other groups of atoms or molecules opened up a new world of synthetic chemistry and the possibility of making a limitless range of new substances.

Medicines, perfumes and plastics from coal

Aside from the production of dyes, this surge in interest in synthetic organic chemistry also brought benefits to medicine through the work of the German industry. Liebig had shown that chloroform (trichloromethane $CHCl_3$) could be made by reacting chlorine with acetone, and this became widely used as an anaesthetic after it was first tried on humans by James Young Simpson in Edinburgh in 1847. Its use in childbirth became acceptable when John Snow, the doctor who found the cause of cholera, gave it to Queen Victoria at Prince Albert's request for her eighth delivery in 1853 – prior to this it was held by the Churches that women were ordained by the Bible to suffer during childbirth. Phenolic disinfectants found a wide application after Joseph Lister had shown the value of antiseptic surgery in 1867. Aspirin, paracetamol and other drugs to reduce temperature and pain, important and common symptoms resulting from inflammation in the body, were to follow.

After Louis Pasteur's discovery of the importance of micro-organisms in disease, Robert Koch (1843–1910), then a country doctor in Germany, used synthetic dyes to colour bacteria so he could study them under the microscope, and went on to discover the one that caused tuberculosis. He hoped in vain that he might find a dye to kill the germs, but his assistant, Paul Ehrlich (1854–1915), took up the search. Encouraged by reports of studies in which an organic arsenic compound had been used with success against the tropical disease sleeping sickness, which is caused by a parasite called a trypanosome, he and his assistant, Sahachiro Hata, systematically studied hundreds of synthetic organic arsenic compounds. In 1909 number 606 proved effective not only against the trypanosome but also against the syphilis organism. This he called Salvarsan; it was effective in human cases of the diseases and thus was the first synthetic chemotherapeutic agent.[5]

There was then a long gap until in 1934 Gerhard Domagk, returning to the investigation of organic dyes, found one called Prontosil red (see Fig. 4.1), which was effective against the very dangerous bacterium *Streptococcus* that caused septicaemia. The active component, sulphanilamide, became famous for curing childbirth fever, a highly fatal condition, and also for curing Winston Churchill of a life-threatening pneumonia during the Second World War. It was widely marketed as M&B, named after the manufacturer, May and Baker.

Not only was coal tar a source of the original chemotherapeutic agents, but the tar itself was also found useful in treating skin diseases. It had proved to be responsible for the success of the great twentieth century industries associated with organic chemistry and pharmacology.

Two further products of the organic chemistry industry were perfumes and plastics. The name of one family of chemicals, aromatic hydrocarbons, gives the clue to their use for making perfumes. These are all based on the ring structure of benzene described above, manipulated chemically to produce the desired odour. This, of course, survives as a major industry producing both perfumes and flavourings for personal use and consumer materials. The plastics industry also started in the early twentieth century with coal as the raw material. The first plastic to be synthesised was Bakelite, patented in 1909 by the Belgian Leo Baekerlande working in the USA. He made it by condensing together two chemicals produced from coal, phenol and formaldehyde. This also was the start of a huge industry in which molecules of one or other simple organic chemical are joined together (polymerisation) to make large molecules of a durable material – for example, the gas ethylene is used to produce polyethylene (polythene) and the gas vinyl chloride to produce polyvinylchloride (PVC). Nowadays the chemical feedstock is oil rather than coal, but the principle is the same.

Electricity

Coal gas for lighting was eventually displaced by electricity. Problems associated with distribution were solved after Sebastien de Ferranti, a Liverpool-born electrical engineer, designed a high-voltage alternating current supply system based on his power station at Deptford on the bank of the Thames in 1889. Many other power stations followed, all burning coal to power the generators, and these were to make a major contribution to air pollution as the use of electric power spread. In 1929 an Act of Parliament established a Central Electricity Board in the UK. A network of cables, mostly carried by overhead pylons, was built to connect the main power stations and by 1938, just in time for the Second World War, a National Grid had been established in Britain. The availability of electricity slowly transformed my mother's life raising her five children at home, as gaslight had transformed my grandmother's. First, a vacuum cleaner arrived, then a sewing machine and later a washing machine and an electric mixer. All were luxuries when they came; they still would be for many people in the world.

The generation of electricity requires power, but any available source may be used. Early generators therefore used water power or coal, and the latter quickly became the dominant worldwide source, being cheap to produce and abundant in many countries. However, three serious problems associated with widespread use of coal for electricity generation became apparent in the mid-twentieth century: local air pollution from soot and sulphur dioxide, water acidification from long-distance transport of acidic aerosols, and ultimately global climate change from the carbon dioxide produced. This has led to moves to reduce coal burning in Europe and eventually worldwide. Now natural gas, oil and nuclear power are replacing coal in the production of electricity, with increasing contributions from non-fossil sources. Worldwide, however, at the time of writing coal is still a major fuel for electricity production. These issues are discussed further in chapter 12.

Coal production and the workforce

The coal industry in Britain reflected the importance of all these industries, flourishing through the nineteenth and early twentieth centuries, reaching a peak of 258 million tonnes annually before the First World War. Thereafter it went into a slow irregular decline, as cheaper coal became available from elsewhere. It picked up slightly after the Second World War in response to the economic recovery after the damage by bombing and the shipping blockade, but then declined steeply to its virtual closure save for some opencast mines in the early twenty-first century. Employment in the UK coal industry reached its peak at 1.25 milion men in 1920. In contrast, world production of coal rose steadily through the twentieth century and has only recently, from 2010, shown signs of decline as China has cut back on its usage in response to realisation of the facts of climate change.

The reign of King Coal, 1800–2000

The reigns of most sovereigns are marked by both the achievements and setbacks for which we remember them. Historians looking back at King Coal will note the development of the steam engine, steel production, gas light, electric power, and the organic chemical and plastics industries, with the consequential Western Industrial Revolution. But they will also note the costs in human health and life, first to the men, women and children who laboured in the mines, from accidents and diseases, to those working with chemicals that caused cancer, to those affected by air pollution, and ultimately to all of us and the wider ecology of the planet, affected by a changing climate and by its insidious permeation by man-made chemicals and microplastics. It is from these, sometimes unforeseen, adverse consequences of King Coal's reign and that of his successor King Oil that we need to draw lessons. First among these were the lung diseases from which innumerable miners were to suffer. Later came the challenge posed by air pollution from coal burning and later vehicle exhausts, and finally the recognition that our use of coal and other fossil fuels was influencing adversely the climate and ecology of the planet itself. These harmful consequences are the subject of the remaining chapters.

Chapter 5

The lungs and their diseases

The world's attention was drawn anew to the dangers of mining in October 1966, when I was embarking on a specialist career in Liverpool and taking a special interest in how the lungs worked. On the morning of 21 October, in Aberfan, South Wales, a huge mine spoil tip, destabilised by heavy rain, collapsed and ran down the hillside to bury a primary school where the children had just assembled in their classrooms. In the most heart-breaking event of my lifetime, 116 children and 28 adults were killed immediately from the impact or from suffocation, and the nation wept as we were reminded of the terrible toll borne by the families of those who provided us with energy. Two years later, in November 1968, people living around Farmington in West Virginia were woken at 5.30am by a huge explosion. The local coal mine was known to be gassy and poorly maintained. Twenty-one men were eventually saved, hauled up an emergency shaft in coal buckets, but 78 had been either blown up or suffocated. Seventeen bodies were never recovered. Later it was discovered that a ventilation fan was broken but the alarm that would have alerted the men to danger had been disabled.

Breathing is something that we hardly notice until we exert ourselves, but an inability to breathe, suffocation, is one of the most frightening of all sensations. When I researched the treatment of asthma early in my career I saw the fear in the eyes of patients suffering a severe attack and the relief when they responded to treatment. I taught my students that the most dangerous warning sign in such attacks was when the patient stopped fighting for breath, too exhausted to continue; then was the time for a respirator to be used until the medicines had time to work. Early accounts of disease in coal miners stress the importance of breathlessness as a symptom, and efforts to understand the effects of coal dust on health have involved measuring the function of the miners' lungs.

West Virginia

My first meeting with coalminers was in 1969 in West Virginia when the memory of the Farmington disaster was still on everyone's lips. I had just received a fellowship to do research into lung diseases, having jumped over most of the hurdles erected by the British National Health Service on the path trodden by junior doctors before becoming consultants in those days. These included amassing sufficient experience of managing patients in different aspects of hospital medicine, passing the examinations required to become a member of the Royal College of Physicians, doing some part-time research and publishing it in medical journals, and obtaining a medical doctorate by dissertation from that research. One last optional extra, which we jokingly called BTA (been to America for a year's research), was known to help one obtain a post in a teaching hospital. I had taken steps to go with my young family to Africa instead, but a civil war in Nigeria made this impossible, so America it was. West Virginia offered the opportunity of studying coal miners' diseases.

The University of West Virginia was in the town of Morgantown in the north of the State (Fig. 5.1). In those days West Virginia was depressed and extraordinarily behind the times for the USA, apparent even to someone who had experienced the difficulties of living in wartime Liverpool and the austerity of the post-war years, but it gave me the opportunity of meeting many of the people who lived there and had worked in the mines. And coal mining was what West Virginia did. The whole state was sitting on a huge coalfield and the coal was put into barges on the Monongahala and Ohio rivers or onto trains, and transported north to the great city of Pittsburgh where Andrew Carnegie had made his fortune and founded the steel industry. The story is a familiar one to people from mining communities; rich men's companies come into your land, you dig up the coal, and they pay you a pittance and use the coal elsewhere to make more money for themselves.

Figure 5.1 West Virginia University Medical School in 1970.

West Virginia was a land of contrasts. On one hand there were the beautiful Appalachian Mountains, rounded hills covered with forest where the people went hunting, which occupied almost all the state. Then there were the rivers, some of which ran yellow with sulphur run-off from the mines, and the huge open-pit or strip mines with their enormous diggers and trucks. There were the relatively prosperous towns of Morgantown with its university and medical school, and Charleston, the state capital with its college and polluting chemical factories along the Kanawha River[1] running from Gaulley Bridge north to join the Ohio River near Pittsburgh. Then there were the wooden shacks, in hollows between the hills, with their rocking chairs or swings on the verandas where the country folk and miners lived. At Gaulley Bridge was a large hydro-electric scheme, notorious for the fact that during the Great Depression several hundred workers digging the tunnel through the mountain had died over a short period from silicosis, a disease we shall meet in the next chapter. A plaque there celebrated the engineering achievement but made no mention of the deaths.

The miners I met were mostly poor but self-reliant, descendants of settlers who in the trek west in the eighteenth century had stayed in the hills and settled there to live by primitive agriculture and hunting. The original settlers were eighteenth-century Scots/Irish who brought with them the fiddle music and ballads that, together with the Afro-American banjo, form the basis of hillbilly music. The fortunes of the inhabitants and those of their forbears had from the late nineteenth century been subject to the vicissitudes of the market for coal and steel and the power of the coal companies over them. Unionisation had been resisted by the companies and the poor working conditions had, from the first years of the twentieth century up to 1933 when Federal Law forced unionisation, led to frequent strikes. Even intermittent armed struggles between miners and mine owners, supported by private armies, had occurred, and attempts by the owners were made to break strikes by shipping labour from the coalfields of central Europe into West Virginia in trains armed with machine guns.[2] Several of the miners I met had Polish or Slovak surnames. These mine wars had occurred in a state whose origin was in war, the Civil War, when as part of Virginia it seceded with the Southern States, then became the separate State of West Virginia in 1863 when the western counties seceded again from the South. Even after the war ended, counties and families remained divided and feuds, often with guns, persisted well into the twentieth century. The miners had long memories and tales to tell, as I studied the effects of their trade on their lungs.

The lungs and how they work

Of all the organs in the body, the lungs and the heart are the easiest to understand. Described very simply, the lungs supply the body with oxygen and rid it of the waste gas carbon dioxide, just as Lavoisier and his wife described in the late eighteenth century. The right ventricle of the heart pumps blood through the lungs' capillaries where red blood cells pick up the oxygen and release the carbon dioxide, carrying the oxygen back to the left side of the heart. This then pumps the oxygenated blood though arteries round the body to nourish its organs before it returns via veins to the right side, the circulation that William Harvey (1578–1657) first described in 1628. The red blood cells contain a pigment called haemoglobin, which turns red when it combines with oxygen, giving arterial blood its characteristic colour.

When we breathe we suck air into our lungs through a tree-like formation of dividing tubes, the trachea (windpipe), bronchi, and bronchioles, which branch repeatedly and become narrower until they end in millions of tiny bubble-like air spaces called alveoli (Fig. 5.2). Normally these spaces are just visible to the naked eye and have elastic walls that allow them to stretch as air is sucked in by the action of our diaphragms and the muscles between the ribs in our chest wall.

Breathing in is an active process, requiring effort, but breathing out just requires the muscles to relax and the elastic recoil of the lungs does the work. Of course, if we are breathless or want to blow out a candle, we can use our chest muscles to breathe out more forcefully. The gases, oxygen and carbon dioxide, simply diffuse between

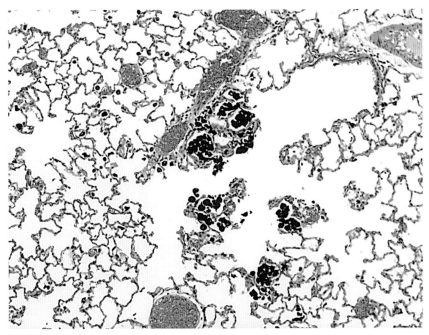

Figure 5.2 Photomicrograph of small airway coming from top right, dividing into terminal airways and multiple alveoli. Dust deposits within macrophages are seen at the centre. Red blood cells mark the small vessels that pick up oxygen from air in alveoli (courtesy of Prof. Ken Donaldson).

the air in the alveoli and the blood through the very thin elastic wall between them. This is driven by the difference in gas pressure in the two compartments, the alveoli and the very fine capillaries that carry blood from artery to vein. This mechanism is roughly the same in all air-breathing animals.

Slightly more complicated is another important function of the lungs, defence from attack by enemies. We have all evolved in a competitive environment, which we tend to see in terms of political or commercial disputes, fights or warfare. However, the main enemies of *Homo sapiens* apart from himself, as of all plants and animals, have always been the organisms causing infectious disease; I have mentioned some of these briefly in chapter 3. We do not always realise it, but the air is home to many micro-organisms; bacteria, fungal spores and even viruses flying in little particles of organic matter. Fortunately, most are harmless but some, such as the germs that cause tuberculosis, influenza or pneumonia are potentially deadly. In fact, they are deadly partly because they have themselves evolved mechanisms that overcome our defences.

Three of our defences are in the lung, which shares with the skin continuous contact with the air. First, the airways from our noses to the smaller bronchi are lined by a membrane that both secretes mucus and also has a mechanism to convey this mucus upwards (or backwards in the nasal cavities) to the throat. Particles including germs caught in this mucus thus get removed from the lungs. This conveyor belt of mucus is operated by the cells lining the airways, which are equipped with microscopic hair-like projections called cilia that move in a coordinated manner like the blades of a Roman galley or a rowing eight, sweeping in one direction then relaxing back (Fig. 5.3). One of many apparent miracles of evolution has resulted in these bronchial lining cells having a mechanism identical to that in the cilia that many micro-organisms use to propel themselves about in the soil.[3]

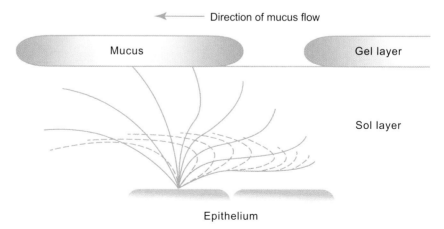

Figure 5.3 Sketch of the action of cilia, acting in a liquid layer to move mucus floating on the surface upwards from the airways (from a draft by author).

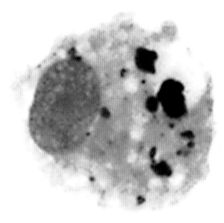

Figure 5.4 A lung macrophage engulfing particles of dust (courtesy of Prof. Ken Donaldson).

The second defence mechanism is provided by another cell that also mimics many micro-organisms, the macrophage (the word literally means big eater) (Fig. 5.4). This cell is manufactured in the marrow of our bones and passes into the blood before becoming lodged in various organs, including the lungs. There one of its jobs is to swallow small organisms like bacteria and destroy them or carry them away, either up the airways or through special narrow drainage tubes called lymphatics that run alongside blood vessels and lead to lymph nodes. Both in moving and in swallowing and destroying germs or particles, the macrophage behaves like an amoeba, using the same internal mechanisms, to stretch out limb-like projections or pseudopodia (false feet). This macrophage also plays an important role in the third important defence mechanism, which is the immune system. In very simple terms, the macrophage sends out chemical signals that attract other cells from the blood to fight infection by a process of inflammation.[4] This is the battle familiar to all of us when we get an infected spot or boil; there is swelling, heat, redness and pain, all of which resolve, sometimes leaving a scar, when the battle is won with or without the help of antibiotics. On the other hand, if the defences fail, the infecting germs spread to local tissue and via the blood to cause septicaemia, fever and sometimes death. To understand lung diseases, it is necessary to know something of these few basic mechanisms.

How the lung defends itself from infection

Throughout the history of *Homo sapiens* the main danger to the lungs has come from infection, and this remains the most important danger to this day. Infection is part of the evolutionary struggle for survival between competing species that all life is engaged in on the planet. Bacteria survive to reproduce themselves by finding a suitable substrate, something which provides them with the nourishment they need to survive and reproduce, just as our remote ancestors sought suitable land on which to grow their crops or hunt their prey. Although generally we live in harmony with our own resident population of micro-organisms on our skin and in our gut, unfortunately some more dangerous germs see us as suitable for their reproductive purposes. We call them pathogens and their effects on our health have evolved so that their many offspring might be passed on to others by cough or touch.

In the lung, infection is familiar to us as bronchitis, characterised by cough, fever and production of green sputum, or as pneumonia with fever, breathlessness and sometimes chest pain; the former is due to infection of the bronchial tubes and the latter to infection of the lung air spaces. Both of these conditions occur acutely in response to viral or bacterial infections and are usually either self-limiting or respond to appropriate antibiotics. Longer-lasting infections, wherein the body's defences hold the infection at bay but do not eradicate it, also occur and are characterised as chronic. The one infection that has had the greatest influence on mankind's history, greater even than plague or malaria, is tuberculosis, a disease that acquired the description 'The captain of all these men of death'.[5] It is caused by a bacterium (*Mycobacterium tuberculosis*) that normally lives in the soil but, as a defence against being eaten by amoeba-like creatures, evolved a mechanism for living and multiplying within cells. This allows it to turn the tables on our defence mechanisms, multiply within macrophages in our lungs, and eventually to cause a persistent (chronic) infection and inflammation that destroys the lung tissue or spreads through the body to cause death.

It is of interest to note that the first effective medicine discovered to treat tuberculosis was found by soil biologists who were investigating other soil microbes; they understood that some would exist naturally that were able to compete with the TB germ in its environment.[6] This was a logical extension of the serendipitous discovery of penicillin by Alexander Fleming, when he found the streptococcus bacterium to be inhibited by the fungus *Penicillium*. Eventually these biologists, principally Selman Waksman and Albert Schatz, found a grey fungus, *Streptomyces griseus*, which antagonised the growth of *M. tuberculosis*. From this was extracted the agent called streptomycin, the first anti-tuberculosis antibiotic. In evolutionary terms, the ability of the tuberculosis bacterium to cause chronic, slowly progressive disease is beneficial to it, as the victim coughs and over a prolonged period spreads it to those close by, ensuring its survival.

The analogy between infectious agents and particles of dust

The above defence mechanisms are shared by all air-breathing animals and have evolved to recognise and remove or neutralise the minute organisms that are able to penetrate to the part of the respiratory tract where oxygen and carbon dioxide are exchanged. When mankind first lit fires in his huts or caves he was also subjected to particles derived from combustion in the air. Because of the design of the bronchial system, only very small particles can reach this critical area, the larger ones falling onto the walls of the bronchi as the air passes into the deep lung. Once on the bronchial wall, any particles are conveyed up to the throat in the mucus to be swallowed or coughed up. The critical size allowing particles to reach the alveoli is around 7 to 10 micrometres;[7] some of these are deposited on the alveolar walls and the rest remain suspended and are exhaled in the breath.

A fire necessarily produces particles of soot of a wide size range, from visible particles of several millimetres down to nanometres.[8] When smoke is inhaled the soot particles are treated by the lung as if they were bacteria. Thus we rely on our bronchi to remove the larger ones and our macrophages to remove the smaller ones. Whether they harm us, by causing inflammation of the bronchial tubes or around alveoli, depends on how effective these mechanisms are and also on the inherent toxicity of the particles. As we shall see, the defence mechanisms are very relevant to understanding the effects of inhalation of dust and fumes as well as of bacteria.

Investigating disease – clinical and epidemiological methods

Understanding disease in general involves two separate but complimentary methods of both investigation and logic: investigating the individual and investigating groups of people. The first of these, clinical investigation (the word clinical comes from the Greek for a bed), should lead to making a diagnosis, that is putting a label on the disease a person suffers from. This in turn leads to the application of current knowledge in order to estimate a prognosis (what is likely to happen) and decide on the best treatment. This approach depends on *deductive logic*, using accepted knowledge to reach a conclusion, arguing from general knowledge to the particular problem of the individual. It is a process that we are all familiar with from visits to our doctor and from reading Sherlock Holmes stories. The second method, study of the health of groups of people in their environment, is what is called epidemiology (from the Greek for study among the people, a term introduced by Hippocrates). This measures the amount of health and disease or some aspects of them in populations, in relation to various factors in the environment that may act to influence them. Arguing from the particular findings to the general, *inductive logic*, the outcome should be a description of the association of the disease or condition with various risk factors and should lead to preventive measures, as with Budd's use of disinfectants or Snow's actions leading to improved management of sewage.

A good illustration of application of the two methods comes from the well-known story of lung cancer. Although this was known to medicine as an uncommon disease from the eighteenth century when diseases started to be understood in pathological terms, it was noticed to be becoming much more common by the 1930s, coincident with the increase in cars on the roads. Clinical doctors diagnosed it by chest radiographs and a few lucky patients were cured by surgeons who removed the affected lung. Application of clinical deductive logic led to its being diagnosed more easily, but cure was rare and hard to achieve, more than 90% of patients dying within a year or two of diagnosis. It became important to find out what was causing the apparent epidemic. While exhaust smoke from cars was suspected, early studies of groups of patients with lung cancer and comparison of their lifestyles with those of people without cancer or with other diseases led to suspicion falling on cigarettes, use of which had increased steadily in men since the 1914–18 war, and in women following the 1939–45 war. Further larger epidemiological studies supported this finding and the evidence gradually increased, though the argument that smoking actually caused the cancer was strongly opposed by vested interests.

Eventually, it was shown that the risk of lung cancer was related to the length of time people had smoked, and that it was lower in people who had ceased smoking. It took decades before this was accepted by those who profited from the tobacco industry, but public health action in the UK and USA persuaded more and more people to stop smoking, and the numbers of people dying from lung cancer had started to decrease in men in Britain by the end of the twentieth century. By then, few argued that smoking was not responsible, and the success of action based on the evidence of association was apparent. Surgeons were able to claim credit for individual lives saved, while epidemiologists could claim whole populations were being prevented from getting the disease.

Methods of investigation

The tools used by the clinical doctor and the epidemiologist are basically the same, although the epidemiological ones need to be simplified and formalised to enable studies of large numbers of people. These tools include: questions about symptoms; examination including investigations of blood, sputum, radiology and lung function in life, or through autopsy or death records. In the history of the discoveries relating to dust diseases of the lung, the pneumoconioses recounted in the next chapter, it will be noted that early accounts of the diseases are clinical or pathological ones, and that these are often accompanied by speculations about cause; these speculations vary over time and have been related to the current ideas about disease causation in general.

Until the detailed description of the human body, *De Humani Corporis Fabrica* (On the structure of the human body) by Andreas Vesalius (1614–64), a Flemish doctor working in Padua, and the description of the appearances of diseased organs, *De Sedibus et Causis Morborum* (On the seats and causes of diseases) by Giovanni Battista Morgagni (1682–1771) also in Padua, post mortem examination was rarely used to discover what had led to people's deaths. Notable scientists such as Leonardo da Vinci (1452–1519) and William Harvey had used dissection (for example, of animals and executed criminals) to investigate the workings of the body, but the performance of an autopsy routinely after patients had died was only popularised by physicians in the nineteenth century, notably René Théophile Laënnec (1781–1826) in Paris. Laënnec's book on the use of his invention, the stethoscope (Fig. 5.5),[9] and his fame as a teacher influenced physicians across the world in careful physical examination of their patients and in performing a final audit by necropsy if the patient died, leading to a great increase in medical understanding of disease.[10]

Two other important advances in medicine that came after the introduction of the stethoscope enabled doctors to learn more about lung disease in life, rather than waiting for the patient to die. First, following Pasteur's work on bacteria, Robert Koch (1843–1910) in Germany discovered *Mycobacterium tuberculosis* in 1882, and it became possible to identify this and indeed many other germs in

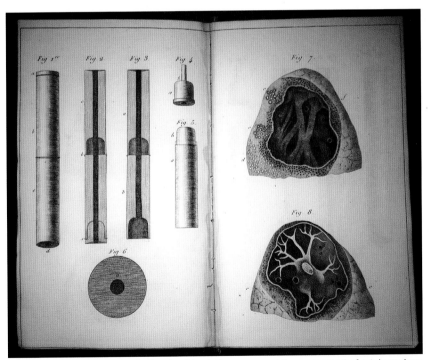

Figure 5.5 Laënnec's book showing his stethoscope and the upper part of a tuberculous cavitated lung. In the text he describes the breath sounds heard through a stethoscope held over the chest of a patient with such a disease. From R.T.H. Laënnec, *Traité de l'auscultation mediate*, 1837 edition.

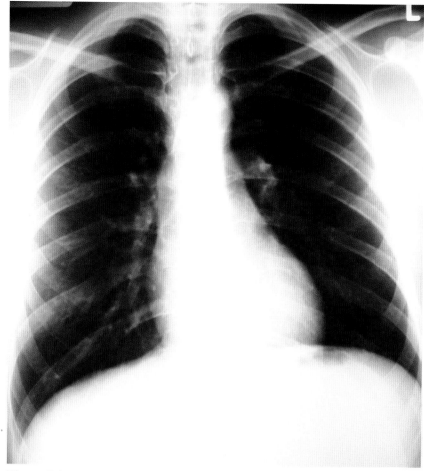

Figure 5.6 A normal chest radiograph showing the shadows of the heart in the centre, diaphragms below, and the ribs and collar bones. The lungs appear black and you can make out blood vessels going from the heart into the lungs.

patients' sputum and grow them in culture media, thus helping to remove the clinical confusion between tuberculosis and mineral dust diseases, as discussed in the next chapters. Secondly, William Roentgen (1845–1923) discovered X-rays in 1896 and their progressive application led eventually to their use in clinical diagnosis from the 1920s onwards (Fig. 5.6).

Finally, some practical methods of assessing the functions of the lung enabled doctors to obtain a quantitative measure of a patient's disablement, using a spirometer. This instrument, initially a simple gasometer, similar to those used by Lavoisier and Thackrah previously, with a calibrated bell and a tube leading to it, enabled measurement of the amount of air that could be expired in a single maximum breath, the so called vital capacity. It was introduced to medicine by

Figure 5.7 A modern portable electronic spirometer.

a Newcastle-born physician working in London, John Hutchinson (1811–51) in 1846. Hutchinson studied many individuals and showed, among other things, the relationship of vital capacity to height. His spirometer was later developed to incorporate a timer so as to measure the rate of expiration. Subsequently, portable electronic methods were developed in the mid-twentieth century (Fig. 5.7).

Further advances in the 1950s made it possible to measure the total volume of air in the lungs, the rate at which gases passed from lung to blood, and the mechanical properties of the lung. From the point of view of this book, the most important lung function test is referred to as spirometry, a forced expiration following a maximum inspiration which allows the vital capacity (VC), the amount blown out in the first second (forced expiratory volume in 1 second, FEV_1), and peak expiratory flow rate to be measured (Fig. 5.8). These serve as a practicable and simple means of assessing airways obstruction for use in population surveys as well as in the clinic.

Both chest radiographs and lung function testing were to play an important role not only in clinical diagnosis but also in epidemiological studies in the community, using various strategies formulated over the twentieth century. The development of these methods allowed description of several separate features of lung diseases:

- First, from pathological description of the appearances of the lung post mortem, a diagnosis (or label);
- Second, by lung function testing, a means of describing the disablement associated with the diagnosis;
- Thirdly, by chest radiology, a means of determining the presence and some characteristics of the disease in life; and

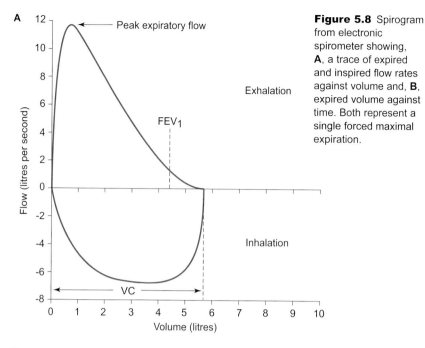

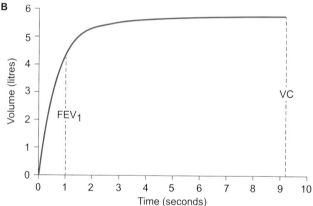

Figure 5.8 Spirogram from electronic spirometer showing, **A**, a trace of expired and inspired flow rates against volume and, **B**, expired volume against time. Both represent a single forced maximal expiration.

- Fourthly, from population studies, a description of the prevalence (frequency in a population at any one time) of the disease, its incidence (rate of occurrence over time) and its effects on longevity.

Ultimately, it has become possible to determine the risks of getting diseases in relation to exposure to hazards in the environment.

The common lung diseases

In order to understand the story of coal miners' lung diseases it is helpful to know something of the other, more common, diseases that working men would have been prone to at the time when doctors first started to notice lung problems in coal miners. In the eighteenth and nineteenth centuries, all that doctors had to lead them to a diagnosis was the ability to ask questions about symptoms and to make a physical examination, using only the senses of sight, touch, hearing, and sometimes smell. The only test of the accuracy of diagnosis would have been an autopsy, so it is likely that few doctors ever found out whether they had been right or wrong.

The process of diagnosis was made more difficult as many symptoms are common to different diseases. For example, cough and breathlessness are features of most lung and heart diseases, and weight loss is frequent in most serious long-term illnesses. Four conditions in particular will have caused diagnostic problems for doctors until the advent of microbiology, chest radiography and lung function testing in the early twentieth century. These are asthma, tuberculosis, heart failure and chronic obstructive pulmonary disease (COPD).

Asthma

We now understand asthma to be a condition of intermittent attacks of wheezy breathlessness that may start and remit at any age, but that sometimes becomes persistent and disabling, and may even be fatal during severe attacks. It is known to have multiple genetic and environmental causes, both in the sense of its initiation and of provocation of attacks, but much of this knowledge is of relatively recent origin. The original meaning of the term 'asthma' was from the Greek for breathlessness, and older accounts often refer to breathlessness from many causes as asthma, for example miner's asthma. Restriction of the use of the term to what we now think of as asthma started in the nineteenth century, but the first clear account by someone who suffered from the condition himself was by the physician Sir John Floyer in 1698. He had kept a diary of his symptoms and described their spasmodic nature and variability in response to various factors. Another doctor, Henry Hyde Salter, similarly wrote a detailed account of his own condition in 1873, and by the early twentieth century asthma in the form we understand it today was described in medical textbooks, although the term also persisted up to the 1960s, as cardiac asthma, to describe acute heart failure.

The main physiological feature of asthma is intermittent reduction of the diameter of the lung's airways owing to three mechanisms: contraction of muscle in their walls, an accompanying inflammation and swelling in their lining mucous membrane, and the secretion of sticky mucus into the narrowed airways. As the airways narrow, so the airflow through them is progressively impeded, and this is felt as difficulty in breathing in, usually also with a wheeze as the air is let out. This abnormality became measurable as the amount of air that could be blown out in the first second of a forced expiration after a full breath in, the so-called forced expiratory volume in one second (FEV$_1$, see Fig. 5.8), after the invention of the timed spirometer. However, this was not used routinely in medicine until well into the twentieth century. Thus throughout the nineteenth and early twentieth centuries there was an understandable confusion between asthma and other causes of breathlessness, especially in middle-aged and older people, in whom true asthma was wrongly thought to be relatively uncommon.

Tuberculosis

Tuberculosis (TB) was one of the infectious diseases responsible for most premature mortality in Britain in the seventeenth to early twentieth centuries. Before the discoveries of Louis Pasteur and Robert Koch in bacteriology, these diseases were thought to be due to a form of fermentation, and known as zymotic conditions. But TB was and still is a worldwide plague, killing more young people than any other disease. It commonly started as a mild infection in children, who often seemed to recover, but then a few years later relapsed and the victim slowly wasted away with a persistent cough and increasing breathlessness – in each case a family tragedy familiar to anyone who has seen Verdi's *La Traviata* or Puccini's *La Bohème*. It acquired a romantic image that was much made use of by poets, novelists and musicians, on account of the tragic circumstances of, especially, young people becoming pale and thin and dying prematurely. Scientists even proposed theoretical explanations of its ability to remove inhibitions and release genius, such as by lowering the oxygen supply to the brain.[11] However, the reality was misery, penury and premature death, particularly among the poorer people living in overcrowded slums. So common did TB become in the conditions of the Industrial Revolution that it was responsible for about a quarter of all deaths in Britain's town and cities.

Two synonyms for TB implied wasting away – phthisis and consumption. Up until the widespread availability of X-rays from the 1930s it was impossible to diagnose the disease at an early stage, so usually it was only recognised when it had already done considerable damage to the lungs. At this stage the mortality was 50%, even in patients treated by isolation in sanatoria and exposure to fresh air and sunlight which, with occasional surgical interventions, was all that could be offered. A cure, by combining three effective antibacterial drugs, became available in the 1950s and this, together with the above public health measures which had already made an impact from the start of the twentieth century, reduced the prevalence of the disease greatly in the industrialised West.[12] However, it persists as a major killer across much of the rest of the poor world. Bacterial resistance, from failure to apply appropriate treatment and lack of effective public health measures, are now allowing it to creep back and spread in a severe, antibiotic-resistant form, as individuals and populations travel and migrate.

During the Industrial Revolution, TB was the main killer in Europe, and it would have been a fair bet that anyone who had a persistent cough and lost weight would be suffering from it. It is not surprising, therefore, that a coalminer with advanced lung disease due to coal, now called pneumoconiosis, would be told that he had phthisis or consumption and that, even when the pathology of the condition became recognised at autopsy, the disease was labelled miner's phthisis. Further, in the absence of knowledge of the causes of the diseases, there would be argument as to whether coal mining could cause phthisis, or phthisis made people more susceptible to the effects of coal. This confusion continued well into the twentieth century, particularly when in South Africa a lung disease caused by inhalation of stone dust, silicosis, was found to be occurring in populations of gold miners. These unfortunate men also lived in circumstances that encouraged the spread of TB, and it became apparent that the presence of silicosis made them more susceptible to this infection. As we shall see, even up to the 1960s, it was still being argued that TB was in part responsible for the harmful effects of coal on miners' lungs.

Chronic bronchitis and emphysema (COPD)

As stated in the introduction, in the mid-twentieth century one of the most common reasons for admission to hospital in Britain was shortness of breath and cough, often with associated heart failure, from a condition then known as chronic bronchitis. Initially the patient had suffered recurrent episodes of cough and sputum production in the winter months and, as these became more severe over several years, he (and they were almost exclusively male in those days) became short of breath on exertion, ending up with severe breathlessness even at rest. The usual end point was admission to hospital following a winter viral infection and death from a combination of heart and lung failure. At autopsy the lungs showed varying degrees of what is called emphysema, a condition in which the small air spaces (alveoli) become enlarged from breakage of their walls. The consequences of this damage are threefold: first, the lungs lose elasticity so that the airways are less well supported on expiration, and thus they tend to collapse and breathing out becomes wheezy. Secondly, as the lungs become overinflated it requires more effort to breathe in. You can test this yourself by almost filling your lungs and then taking small breaths in and out at this lung volume. Thirdly, the loss of airspace walls and their associated blood capillaries means that less oxygen can get into the blood, so the patients tend to get blue and breathless on effort.[13] Ultimately the heart, having to work harder to pump blood through the damaged lungs and having insufficient oxygen, starts to fail.

Put simply, the recurrent cough and sputum are the consequence of the bronchitis, inflammation of the airways, and the breathlessness is largely the consequence of the emphysema. The two are almost always combined to different degree and it is convenient to call them by one name, chronic obstructive pulmonary disease or COPD. This, leading to recurrent admission to hospital and ultimately death, was and remains the end result of a slowly progressive process over many years. It is now known to be largely a consequence of cigarette smoking, explaining why women were initially less at risk, since until the 1940s relatively few smoked. Recurrent chest infections, air pollution including workplace exposure to dusts and gases, and genetic susceptibility also played a part in its causation. There had been much medical discussion about the relationships of disease of the airways (bronchitis) and of the air spaces (emphysema) and of the different types of emphysema, but the two central features of the disabling condition are impairment of airflow and destruction of walls of alveoli. The best explanation of its pathology is inflammation caused by inhaled particles from, usually, cigarette smoke causing accumulation of white blood cells that release chemicals that both break down the elastic tissue in the alveolar walls and also damage the lining of the airways.

COPD certainly occurred in the nineteenth century (Dr Samuel Johnson was a famous sufferer and a physician, Matthew Baillie, described the pathology of his lungs in a book published in 1793) but was less common before smoking of manufactured cigarettes became almost universal among men during the First World War, and in women in factories during the Second World War. Indeed, some of my teachers, who had graduated in the inter-war years, told me that emphysema never occurred in women. It soon became apparent that this was wrong, as the disease started to become common among female smokers in the 1970s. In most cases COPD is preceded by years of cough and production of sputum (chronic bronchitis) and, until the arrival of X-rays and lung function testing, the consequential emphysema could rarely be diagnosed in life. Early accounts make clear that miners suffered from bronchitis, but except in rare cases the black appearance of their lungs was the dominant lesion, and any emphysema was probably disregarded or missed.

Heart failure

A fourth common condition that causes breathlessness and that would have caused diagnostic confusion among miners was heart failure. As mentioned, this condition may be the end result of chronic lung disease of any sort, but also more commonly occurs as a result of disease of the heart itself, usually from reduced blood supply though the coronary arteries or from high blood pressure. Apart from causing shortness of breath on exertion and at rest, it may cause accumulation of fluid in the lower legs (oedema). Until the late nineteenth century it was known as dropsy, and was one of very few diseases for which there was a specific treatment, digitalis. This was an extract of foxglove, discovered and investigated by a doctor in the English midlands, William Withering (1741–99). He published *An Account of the Foxglove and some of its medicinal uses, etc.* in 1786; the drug and its synthetic derivatives have remained in use from that date until today.

The questions about miners' lung diseases

It may be seen that in the Industrial Revolution the working man would have been confronted by a number of serious threats to his health, mostly common to all men and women, but some specific to his trade. The disputes concerning the effects of coal on miners' lungs that I shall describe arose largely from the difficulties confronting doctors in deciding which of these diseases was responsible for a miner's illness. Indeed, whether coal mining itself caused men to be more susceptible to these diseases or even caused some of them became a matter of dispute. In particular, the question as to whether coal mining could cause bronchitis and emphysema was to become the most contentious issue surrounding occupational lung diseases in the mid-twentieth century.

When I arrived in West Virginia in 1969 there were several important unanswered questions with respect to coal miners' lung diseases. On a wave of public sympathy, in response partly to the Farmington disaster, the US Federal Government had enacted a law, the Coal Mines Health and Safety Act, widely known as the Black Lung Act, to compensate disabled miners. There was, however, considerable dispute as to how much disability was attributable to dust exposure as opposed to cigarette smoking, and also what sort of functional impairment was attributable to dust disease. There was also little information nationally on the prevalence of the diseases in the different US coal mining areas. The Federal Government was therefore funding epidemiological and clinical studies to investigate these problems, and I joined a team led by Keith Morgan (1929–2007) and his colleague LeRoy Lapp engaged in this activity, using the methods described above (Fig. 5.9). The aims were to describe the prevalence

Figure 5.9 The mobile unit, containing X-ray, spirometry and clinical facilities, visiting a mine in West Virginia in 1970.

of dust diseases throughout the United States, to identify the conditions in which they occurred, and to investigate the disablement associated with them. With associated work on dust control, it was intended that steps would be taken to reduce risks among miners. The work was based on methods that had been developed by the Medical Research Council and its Pneumoconiosis Research Unit in Britain, as recounted in the following chapters.

I spent two years learning lung medicine, epidemiology and physiology from LeRoy Lapp and from Keith Morgan. We compared lung function in non-smoking miners, with and without the dust-induced disease, pneumoconiosis, with that in non-miners. We also worked on standardising X-ray interpretation, that is, testing rules intended to increase the likelihood that multiple people reading a chest radiograph would record more or less the same amount of disease in it. An interest in the way in which we react to our environment, especially the workplace environment, was to become the dominant theme of my life from this point on. This period of learning was, however, to be followed by an interval when I was to resume my career as a clinical doctor; I returned to the UK as a chest physician on a pathway that eventually took me back to studying coal miners and the disease they suffered from, coal workers' pneumoconiosis.

Chapter 6

Anthracosis – the disease that disappeared

Pneumoconiosis is a name derived from an even longer one introduced by a German pathologist, F.A. Zenker in 1866 (pneumonokoniosis) as a generic term to describe diseases of the lung caused by inhalation of dust. It literally means, from the Greek, dust lung. The first such disease to be described was silicosis from inhaling crystalline silica, to be followed as time passed by recognition of anthracosis, now called coal workers' pneumoconiosis (CWP), from coal dust, siderosis from iron oxide dust, asbestosis, and so on. But the recognition of dust diseases goes back much further into history. Indeed, as mentioned above, such conditions were recorded by Agricola (see chapter 3) and his contemporary Theophrastus Bombastus von Hohenheim, who became known as Paracelsus (1493–1541). Paracelsus was a rebellious doctor who was always in trouble with the authorities. He nevertheless made his name as a teacher and as the discoverer of a partially effective cure for syphilis – small doses of mercury. He is responsible for first pointing out that most chemicals are poisonous in high dose, and that the dose is critical in making something a poison, a concept that is now known as dose-response and is central to the modern sciences of toxicology and epidemiology.[1] Both he and Agricola were aware of the premature mortality from wasting disease in metal miners in the Erz mountains of Bohemia and Saxony, now the border between the Czech Republic and Germany. However, the first clear pathological descriptions of miners' diseases are from the early Industrial Revolution, starting with the disease that became known as silicosis. In this chapter I shall explain the early recognition of diseases caused by dust, why it was difficult for doctors of the day to diagnose them and distinguish them from other common diseases, and how this led to arguments about the existence of a specific disease caused by coal dust.

Silicosis: the first pneumoconiosis

Most stone other than pure limestone contains a crystalline mineral called silica, a combination of the elements silicon and oxygen (SiO_2). There are several crystalline forms, but by far the most common is called quartz. Anyone who cuts, polishes, or drills stone, or otherwise is exposed to the dust generated from it, is thus liable to breathe in particles of quartz. As stone is a very useful material, over the generations ever since men first knapped flint to make axes and arrowheads a very large number of people have inhaled dust containing quartz, and many continue to do so.

In 1796, James Johnstone, a physician from Worcester in the Midlands of England, gave an address to the Medical Society of London that was intended to inform his colleagues of a scandal, and by which he hoped to initiate reforms.[2] Speaking of the needle-pointing industry in what became known as the Black Country, he said:

> Persons employed in pointing the needles by dry grinding them are constantly very soon affected with pulmonary complaints, such as cough, purulent or bloody expectoration; and being so affected, they gradually waste in flesh and strength, and hardly ever attain the age of 40 years. As the business is known to be constantly attended by such fatal effects, the manufacturers find it not very easy to engage persons to work at it; and they who are engaged are so well paid as to get money enough to misspend in drink; being for the most part, in this respect, persons of very irregular manners.

He went on to suggest some protective measures including:

> Nor would it be difficult to contrive a crape hood or gauze helmet, to receive the head and rest on the shoulders, which would prevent a great deal of the metallic and stony particles of dust, which fly off in the operation of dry grinding the needles, from entering the ramifications of the arteria trachea cells of the lungs in the action of inspiration. The cause of the pulmonary phthisis peculiar to the people who follow this business is undoubtedly the continual irritation of the lungs by the dust of small particles of iron and stone and their gradual congestion into small concretions on the air cells of the lung.

This was one of the first descriptions of what became known as silicosis. Johnstone not only drew attention to the symptoms and the fatal course of the disease, but also proposed preventive measures and explained the cause of the pathological features. But, in the fashion of the times, he was unable to resist attributing some blame to the workers' intemperate habits.

Johnstone was not alone in noticing this dreadful consequence of working in dusty industries, which stood out particularly since the wasting disease afflicted previously healthy men who had survived into middle age, as opposed to the more usual consumption (later shown to be tuberculosis) that afflicted particularly the young. This point was picked up by W.F. Alison, who was professor of medicine in Edinburgh in 1823 when he reviewed what he called the scrofulous diseases.[3, 4] Writing of stonemasons, he said:

> I have witnessed many melancholy examples of the disease among them, at the age of 40 or more, and in well-made men of apparently vigorous constitution, … and I have reason to believe that there is hardly an

instance of a mason regularly employed in hewing stones in Edinburgh living free from phthisical symptoms to the age of 50.

By this time the risks of any trade involving grinding or cutting stone were being recognised, especially in France, which was the source of particularly hard sandstone for making grinding wheels.

The consequences of working the French stone were described in England by Thomas Peacock, a physician in London, in an article written in 1860.[5] He commented on the reports of disease amongst French workers and the account by Johnstone above, and reported multiple cases of the same consumption among men working the same stone to produce millstones. He recorded the comments of experienced workers on the differing harmfulness of different types of stone and, very unusually for the time, he devised a controlled study, comparing the health of the stonecutters with that of comparable workers engaged in basket making. This led him to conclude that intemperate habits did not fully explain the disease, a conclusion supported by his finding of particles of dust identical to that in the air of the workplace in the lungs of one of his deceased patients.

In Britain this disease became known as silicosis after Edward Greenhow in 1865 published his analysis of the minerals in a knife grinder's lung, showing it to contain tiny particles of quartz or crystalline silica.[6] By then it was realised that this was the primary cause. Greenhow also showed that the disease occurred in potters, and further reports, extending up to the present time, have shown that the disease occurs in workers cutting roadways in coalmines, crushing and carving rock, grinding knives and swords in the knife-making trade in Sheffield, using powdered silica in the potteries, in sandblasting, and indeed anyone working with, cutting, or drilling stone. Early in the twentieth century silicosis became one of the first group of industrial diseases, as opposed to accidents, that became compensable by law in Britain. Sadly, the disease still occasionally occurs among these and other less expected groups unless appropriate precautions are taken to prevent it. Indeed, in recent years a fatal outbreak has been described in Turkish workers sandblasting denim jeans.

What causes black lungs? The first controversy

As medical students in Liverpool in the early 1960s we attended daily demonstrations in the post mortem room in order to learn about the structural changes in the body caused by disease, the science of pathology. We quickly learnt that the lung retained evidence of the environment in which the late patient had lived. Apart from gross damage caused, for example, by smoking, we saw black pigmentation looking like a fine network on the lung's surface and throughout its substance when it was cut across. We also noticed dense black pigmentation in the lymph glands around the airways where they joined together to form the trachea.

These deposits of soot came from the industrial and domestic pollution that was so prevalent in cities in those days; in contrast, the lungs of deceased country dwellers or of children were pink. This observation had earlier been made by an Edinburgh medical graduate working at St Thomas's Hospital, George Pearson, in a communication to the Royal Society of London in 1813, and published in its Philosophical Transactions of that year.[7] He proposed that the black material was derived from *..sooty matter taken in with the air at respiration and accumulated in proportion to the duration of life.* His experiments showed that the matter in the lungs was like charcoal and stated:

> I think the charcoal in the pulmonary organs is introduced with the air in breathing. In the air it is suspended in invisibly small particles, derived from the burning of coal, wood, and other inflammable materials in common life.

He speculated that these minute particles were more widely present in the atmosphere than was recognised. Obvious as it may seem to us nowadays, this observation was contentious at the time, others proposing that the carbon was derived from carbon dioxide from the exhaled air.[8]

In the early nineteenth century, René Théophile Laënnec, mentioned in chapter 5, was the great authority on lung diseases. He had been born in Quimper in south Brittany in 1781 and by the time of his death 46 years later he had become professor of medicine in Paris, invented the stethoscope, and shown its use in diagnosis. Through his personal teaching and his book, *Traité de l'auscultation médiate*,[9] he had influenced physicians across Europe to understand disease in terms of the changes in the organs of the body. Among his many observations were those on what he called melanosis, a term describing blackening of the tissues. In his book he differentiated a black tumour that often spreads to the lungs, which we now call melanoma, from the black pigmentation described above. He thought the pigment in the former was derived from blood but was equivocal about the cause of the latter. After his early death, ironically from the tuberculosis to which he had devoted so much of his attention, a fourth edition of his book published in 1837 was edited by one of his pupils, Gabriel Andral (1797–1876), professor of pathology in Paris, in which he recorded the report by a doctor in Britain of several cases of black lung in coal miners, as well as adding two cases of his own. This volume included the first known painting of a lung with coal worker's pneumoconiosis, from a 39-year-old man who had worked underground for 22 years (Fig. 6.1).

The paper to which Andral referred was written in 1831 by Dr James Craufurd Gregory (1801–32), a Scottish physician who had also studied under Laënnec in Paris. He reported the case of a coal miner from south of Edinburgh who became breathless and coughed up black sputum, eventually dying of heart failure. An autopsy showed that he had black lungs containing solid masses of black tissue, and

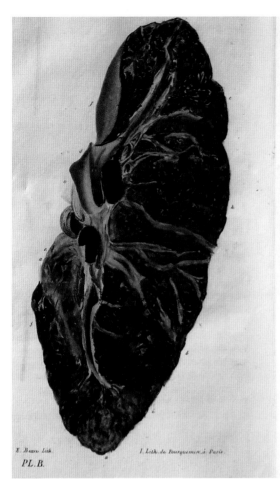

faisai...
de râle muqueux a grus...
térieure de ce côté de la poitrine dont
profondes, un râle crépitant sec à bi
modification aucune de la voix. Au
offrait dans la fosse sus-épineuse le
gauche, ainsi que dans tout le reste
dans la gouttière vertébrale, au niv
plate, on trouvait le bruit de souffl
marquée.

Le malade succomba le même jou
L'autopsie fut faite le lendemain à c
la PL. B ci-contre.) Point de verç
marquée, un peu d'œdème des mall

La cavité gauche de la poitrine p
times et solides, qui occupent les
poumon, et qui sont formées par
fortes; aucune trace de sérosité. Le
adhérences ni fausses membranes.
ont un aspect noirâtre, que la plèv
de traces d'injections sous pleuréale
leur poids est de beaucoup augmen
présente partout une coupe nette.
est noire comme du charbon, et c

PL. B. Coupe du poumon droit vu pa
aa. Points où la plèvre est conservée.
b. Portion du poumon où son tissu est
c. Caverne à parois jaunâtres et épaisse
d. Branche de l'artère pulmonaire.
ee. Ganglions bronchiques entièrement
ff. Bronches ouvertes par la coupe du
g. Lobe supérieur; h, lobe moyen; i

E. Beau Lith. I. Lith. de Fourquemin, à Paris
PL.B.

Figure 6.1 The first known illustration of a coalminer's lung alongside Andral's description of the patient. From R.T.H. Laënnec, *Traité de l'auscultation mediate*, 1837 edition.

Gregory speculated on whether the black material was generated within the body from blood or was inhaled. He wrote:

> The question here immediately presented itself, whether this ought to be considered as a case of infiltration of the substance of the lungs by the peculiar matter of melanosis, or whether the black colour of these organs depended merely upon the habitual inhalation of a quantity of the coal-dust with which the atmosphere of a coal-mine must be constantly charged...

The tissue was examined by the professor of medical jurisprudence, Robert Christison, who confirmed that this material was coal. The lung, preserved in the museum of the Royal College of Surgeons of Edinburgh, was recently rediscovered and the features of coal workers' pneumoconiosis confirmed (Fig. 6.2).[10, 11]

The paper by Gregory achieved what he intended, by informing colleagues working in coal mining areas of the existence of such a disease. In 1834 William

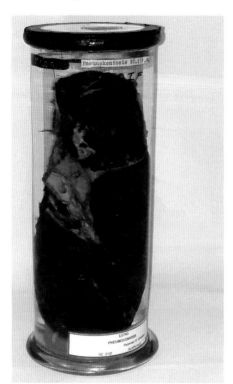

Figure 6.2 The preserved lung of the first man shown to have coalworkers' pneumoconiosis, almost exactly contemporary with the painting in Laennec's book (photo by Prof. Ken Donaldson).

Marshall from Glasgow described other cases and gave them the name 'spurious melanosis', and then 'phthisis melanotica', to differentiate them from what was called true melanosis (the tumour melanoma, as we now know it to be).[12] This first dispute about miners' diseases arose from trying to decide the cause of the black colour found in many lungs at autopsy. Some doctors thought it was from altered blood and others from carbon, perhaps derived from the smoke of oil lamps or from work in coal dust. In the tumour called melanoma they would indeed have found blood, but much later it was shown that this colour was from the pigment melanin in the tumour cells, derived from the pigment cells of the skin where the tumour originated. In a second paper, Marshall described the coal-derived pathology in a series of patients: the progression from a few small black spots in the lung, through aggregation into larger masses, to the development of large, often cavitating masses full of black liquid.[13] He also recognised that some coal miners were involved in driving tunnels, and thus exposed to stone dust rather than coal; in one of his patients he described changes in the lung that would be interpreted today as silicosis. Finally, in 1837 Thomas Stratton published an account of a case of coal miner's lung from the north-east of England for which he coined the term anthracosis.[14] The coal miners' disease had acquired an appropriate name.

Emphysema, the pathological component of COPD discussed in the previous chapter, did not generally feature in discussions of coal miners' lungs in the nineteenth century, but it became important in disputes about the consequences of exposure to coal dust in the twentieth century (see chapter 8). One nineteenth-century contribution is of particular note, however. When a lung is removed from the body at autopsy, its elastic tissue causes it to collapse, so enlarged air spaces are not easily seen. Dr William Craig of Glasgow described a means of examining the *post mortem* lung by inflating and drying it before taking slices.[15] He studied many lungs in this way and was able to show that everyone beyond early childhood had black matter in their lungs, the amount increasing with age, and he was clear that this was inhaled from the polluted air, whether from lamps in the home or from their work. In describing one lung he wrote:

> Its pleural surface presents a greyish or mottled appearance, in consequence of a mixture of the black matter with the dilated air cells. The interlobular lines contain a considerable quantity of black matter. Dilated air cells, varying from the size of a pea to that of a walnut, are seen on its surface … The air cells are greatly dilated throughout its whole substance … In those parts where the air cells are most dilated, the black matter is most abundant …

This was probably the first account of an association between dust in the lung and emphysema. The subject was a 'poor woman of 90 years'; it is likely, given the description of her lungs, that she had worked in a coal mine or sorting coal but this was not stated. In spite of this important observation, inflation of lungs at autopsy only became standard practice following the work of Gough in the 1940s (see chapter 7), and whether coal mining could cause bronchitis and emphysema was to become the most contentious issue surrounding occupational lung diseases in the mid-twentieth century.

In spite of some scepticism in the medical profession, at least one major textbook, published in 1848 by Thomas Watson, professor of medicine at King's College London, gave a description of 'spurious melanosis' in coalminers. He surmised that such a disease had not been recognised before in such a large population of coal miners as were then employed because in their communities there was a reluctance to allow autopsies. This is not implausible, since the early descriptions were from around Edinburgh during the era of grave robbing for anatomical specimens that was occurring there. Indeed, the graveyard at Dalkeith, the mining area where Gregory's patient had worked, contains a watchtower erected to prevent this. The trial of Burke and Hare in 1828, for their notorious shortcut by murdering people to sell their bodies to the anatomist Robert Knox, was widely known at the time.

Does black lung cause symptoms? The second controversy

It would be thought that such clear clinical and pathological descriptions of coal mining disease, and their dissemination in lectures and textbooks across Britain,

Europe and indeed the young USA, would have left no doubt in medical minds that coal mining was a trade that was dangerous to the lungs. Nevertheless, the condition has been associated with fierce medical controversy right through to the end of the twentieth century.[16] Until, at the earliest, the late 1920s, doctors had to rely on their clinical skills to make a diagnosis. They did not have access to X-rays or lung function tests. Miners were as susceptible to tuberculosis, asthma, COPD, and heart failure as other men, so it is easy to understand why confusion of these different diseases, all causing cough, breathlessness and usually loss of weight, should have occurred. Perhaps, when a miner had fallen ill, his symptoms were due to one of these other diseases; was the fact that his lungs had been found to be black at autopsy coincidental?

Looking back to the nineteenth century it is apparent that doctors working in coalmining areas increasingly recognised a disease that could be defined confidently only at autopsy, one that occurred in coal miners and was caused by dust exposure. However, its diagnosis in life remained a matter of speculation, especially against the background of TB and COPD. Only in advanced cases when the patient coughed up copious black sputum, from what we now know to have been breakdown of lung tissue infiltrated with coal dust, could the clinical doctor confidently make the diagnosis. And another difficulty began to be appreciated: some miners who had been perfectly fit from a lung point of view and had lived to a good age were being found at autopsy to have pathological evidence of pneumoconiosis in their lungs. Could it be that coal dust was not toxic and that any symptoms were due to other coincident diseases? Was anthracosis perhaps not a true disease, that is, some pathological abnormality of the body that causes disability, but simply an accumulation of dust in the lung, albeit to a greater degree than occurred in normal town dwellers? Did this disease called anthracosis, pneumoconiosis caused by coal, actually exist?

The role of public health

As noted in chapter 3, the nineteenth century was the era of rising appreciation of public health, driven by the important epidemic diseases and the knowledge that public action was necessary to reduce risks of premature death. This movement had started in France, based on the ideas of the philosopher Adolphe Quetelet, whose book, *Sur l'homme et la développement de ses facultés, essai d'une physique sociale*, advocated the collection of data on fertility, deaths and social factors, introducing the concept of the average person. These so-called vital statistics were first collected in 1821 in Paris by an army surgeon, René Villermé (1782–1863), and showed that the most important determinant of premature death in that city was poverty. This was the start of modern epidemiology, the study of the distribution and determinants of health and disease in populations. In the poorest districts the life expectancy was two-thirds that in the richest. Villermé saw the answer to be in religious education and improvement of morals, and did not believe legislation was the answer, a response that might also find favour with one particular political outlook today. *Laissez faire* did not only apply to economics, but also to social policy. One might comment that circumstances are not very different even today in some cities of the United Kingdom or the United States, other than that the average person now lives much longer than he or she did 200 years ago. In Britain today, life expectancy at birth is about 12 years shorter in the poorest 10% than in the richest 10% of the population. In the UK in 1860 male life expectancy at birth was 40; in 2013 it was 77.

In Britain, William Farr (1807–83), a doctor who had studied medical statistics in Paris, became the first statistician in the UK Registrar General's office, following a law of 1836 making registration of births, marriages and deaths compulsory. He formulated life tables that demonstrated the life expectancies of males and females at different ages according to occupation, wealth and hygiene, and was able to point out the influence of overcrowding and poor sanitation on health and on death from the so-called zymotic, now infectious, diseases. Such tables are used to this day as the foundation of public health policy. The accumulating knowledge from Farr's work was the stimulus to Edwin Chadwick, who as mentioned in chapter 3, had been secretary to the philosopher Jeremy Bentham, to promote the idea of a central public health facility in Britain. His work led to the first Public Health Act of 1848, which set up a General Board of Health for England and Wales. One of this body's reports, written in 1858 by the same Edward Greenhow who was later to make the original observations on silicosis mentioned above, considered, among others, the mining industries of these countries (Scotland had its own system and did not compile occupational data at this stage).[17] He commented:

> There is no class of places in which the influence of occupation on health is more powerful or so evident as in some of the mining districts. Mining operations are frequently pursued in situations naturally salubrious and generally more or less removed from large cities. In many cases the little centres of population that spring up in mining districts are exclusively inhabited by miners and their families, and the classes who minister to the wants of the mining population. Hence the influence of the prevailing occupation on the health of those engaged in it is proportionately evident.

He then went on to tabulate the death rates in the different mining occupations. For example, in lead mining areas childhood mortality from lung diseases was a quarter of that in the populous city of Liverpool but the adult male mortality was 30% greater. Similar trends were apparent in tin and copper miners. However, in English coal mining areas the mortality of both children and adults differed little from, and was in some places rather more favourable than, that in cities or across England generally. Moreover, adult male mortality rates were much lower in coal mining areas than in the metal mining areas save in Welsh mining areas, where a proportion of the population was engaged in mining iron as well as coal, and where there were relatively high death rates from lung disease.

This report was a landmark in pointing to the need for attention to be paid to occupational causes of ill health as well as to its social determinants in cities. However, it may have conveyed a false impression that, in contrast to other forms of mining or industries such as pottery and cutlery manufacture where workers were at risk of silicosis, coal mining was a healthy occupation. In fact it showed that coal mining areas differed in their influences on mortality; in most, male mortality was lower than in areas where other minerals were mined but higher than in cities, whereas childhood mortality and female mortality were generally somewhat lower in coal mining areas than in cities. Importantly, in view of future disputes about the health effects of coal mining, it drew attention to differences between mining areas, South Wales seeming to be particularly unhealthy.

It is apparent that public health data allowed a certain complacency to develop concerning lung disease in coal miners since, as a class, people living in country areas lived longer and had lower risks of TB than those in cities, and comparisons with groups exposed to silica (which not only kills by causing silicosis but also predisposes workers to TB) showed that coal miners were healthier. It also became apparent that there were differences in the risks of coal miners suffering lung disease between different parts of Britain.

Perhaps coal dust is not the problem?
The third controversy

The most influential doctor in the field of occupational disease was Thomas Arlidge (1822–99) who worked in Stoke-on-Trent, the English pottery district made famous by the Wedgewood factory. He had studied the silicosis and lead poisoning that afflicted the workers in that trade and campaigned effectively for improved conditions and regulation. In his book on the Hygiene, Mortality and Diseases of Occupations published in 1892, he wrote:

> There is a widespread belief at the present day that the serious lesions
> of the lungs associated with the calling of coal getters belong to past
> history, or at the most are very uncommon; and no doubt can exist
> that, compared with the past they are becoming rarer thanks to the
> introduction of efficient ventilation, of shortened hours of labour, and
> of the increased attention given to the hygiene of mines.

It may well be that he was correct in assuming a reduction in the incidence of these diseases as the methods of getting coal were improved through the later nineteenth century, but the fact is that the epidemiological methods to justify this belief had not been developed at the time. Nevertheless, in 1906 a Committee on Compensation for Industrial Diseases of the UK Government, having taken evidence from a number of expert doctors, reported:

> We are clearly of the opinion that coal miners are not liable to fibroid phthisis, and although cases of anthracosis, using the term to mean cases in which the lung is charged with coal dust, are commonly met with, we cannot find that in any one that condition has proved to be a contributory cause of death.[18]

By 1908 the Registrar General for England and Wales was able to publish statistics comparing the mortality of coal miners between 1890/92 and 1900/02 in different regions of the countries. These did indeed show a striking reduction in risks of death from all lung diseases over the period, but big differences between regions, the worst being in Lancashire and South Wales. While, save in South Wales, the mortality of coal miners compared favourably with that of all working males across the nations, it was also apparent that in both Lancashire and South Wales coal miners' mortality from bronchitis was double that of all males across the two nations.

In spite of this statistical evidence that something deleterious was going on, the belief that coal mining was relatively harmless to the lungs persisted through the first quarter of the twentieth century, with one proviso relating to exposure to quartz causing silicosis. In 1915 Edgar L. Collis (1870–1957), who had worked as a Medical Inspector of Factories and was an authority on the influence of work on lung health, concluded in his Milroy lectures to the London Royal College of Physicians that:

> I have attempted to justify the claim that dust inhalation plays an important part in determining the occurrence of respiratory diseases. Some dusts such as coal it is true, not only appear to have no power of producing pneumoconiosis, but even may possess some inhibitory influence on phthisis; but most dusts have an injurious influence, and of all dusts that of silica is most injurious.[19]

He recognised that coal miners who were exposed to silica were at risk, while sticking to the belief that coal itself was harmless. He was to change his mind.

Are coal miners' lung diseases caused by coal or silica?

Not long after Collis's lectures, in 1919 silicosis was recognised as a compensable disease in certain specified industries in the UK. These included mines, and it therefore became important to determine whether silicosis occurred in coal miners and was the cause of any disability they suffered. At the same time, interest in simple anthracosis faded, this being regarded as a relatively benign accumulation of dust without serious consequences. In contrast, silicosis was known to be a serious, progressive and often fatal disease of stone workers and knife grinders. In the 1920s, with the use of X-rays especially in South Africa in the new gold mining industry, its harmful effects became widely appreciated. Did coal miners suffer from silicosis, as did other miners?

An important contributor to this debate was Sydney Walter Fisher (1887–1964). Having qualified in medicine in Dublin in 1915 and served in the Medical Corps in the First World War, he worked in general practice in a mining area of South Wales. He became the first UK Medical Inspector of Mines, and later the Principal Inspector. He wrote his doctoral dissertation on silicosis in the mining industry but also had considerable experience in the management of mining disasters and great sympathy for the miners. A clear answer to the question of silicosis in coal mining came in 1924, after the death of one of a group of coal miners in Somerset (in southwest England) who developed a rapidly progressive lung condition. Autopsy showed him to have silicosis. The miners had been drilling through sandstone in order to tunnel to the coal seam. Radiographic studies of these men and a comparable group of miners working the coal showed silicotic fibrosis among the drillers but only what were regarded as minor changes in the controls. Recounting this episode in 1935, Fisher drew attention to this risk among certain groups of coal miners working stone, variously known as drillers, hard-headers, branchers, or brushers in different parts of the UK.[20] Silicosis in specified trades had been recognised as compensable since 1919, and so in 1928 these types of work in coal mines were added to the schedule.

In 1927, Lyle Cummins,[21] professor of tuberculosis in Cardiff and soon to become the premier expert on pneumoconioses (see next chapter), wrote: *Coal miners as a class are nowadays much less liable to respiratory disease than other workers exposed to dust; so much so that coal dust in itself may be regarded as harmless or even beneficial in the concentration encountered in up-to-date mines.* However, having in mind a possible contributory effect of TB, he added later in the article this proviso: *Where any factor exists to hinder its elimination, however, coal dust is capable of accumulating in the lungs and leading to serious or even fatal disease.*

In 1931 a very influential figure had entered into the argument. Edinburgh-born John Scott Haldane (1860–1936) was professor of physiology in Oxford and has to this day a worldwide reputation for his work on the way the body transports oxygen and on the effects of gases, of high altitude and of diving on the body. He had investigated toxic gases in mines, often experimenting on himself. He was responsible for inventing the military gas mask and the mine self-rescuer, and had also introduced the canary to detect gases and the practice of stone dusting to reduce the risk of explosions into coal mines. He had experience of investigating silicosis in tin miners, but believed that coal dust was harmless; his theory, based on mortality statistics, was that bronchitis was the primary disease of coal miners and that the cause of this was the hard muscular work of the men stressing their lungs. He believed that death from silicosis was due to complicating tuberculosis, and therefore since TB was relatively uncommon in coal miners they would rarely die of silicosis. Commenting on the difference between South African gold miners and British coal miners, he stated:

> It has been shown lately that when healthy coal miners are examined with the x-rays, a good many of them present a picture which cannot be distinguished from the picture seen in the case of the Johannesburg miners. If a man in the mines at Johannesburg presented such a picture, it would be regarded as sentence of death, but this picture seems to be found in this country among coal miners who are fairly healthy. I want to make a strong protest against the practice of diagnosing silicosis on mere x-ray examination without knowledge of the man's history and of the kind of dust he breathes.[22]

Haldane was quite right in one important respect. The shadows cast on an X-ray film by inhaled coal and the body's reaction to it look very similar to those cast by silicosis, but the consequences are sharply different, silica causing much more fibrosis and disability. But this did not mean that coal dust was harmless, simply less harmful than silica dust. So by 1930 there was considerable uncertainty in medical minds as to the contributions by different diseases to any disability suffered by coal miners, and indeed also as to the relationship between coal dust exposure and coal miners' relatively good life expectancies.

In summary

Silicosis, anthracosis, bronchitis, tuberculosis: all four diseases occurred in miners, yet as a group they seemed to be as healthy as the generality of labouring men in Britain at the time. What was going on? Initially, in the early nineteenth century the black lungs of coal miners were recognised to be associated with serious disease in Scotland, but it later became apparent that some miners could have black lungs yet remain healthy. It was then established that coal miners could in some circumstances get severe and fatal disease from silicosis, so the tendency was for doctors to assume that disabled miners had this disease and that coal itself was relatively harmless. Miners and those who worked with them were less sure. Noting these disputes, in 1935 a well-known mining engineer, Dr William Cullen, was quoted as saying ... *the mining man could forget about the medical man for the present. His job was to get rid of the dust by every means in his power.* He was right, but the Second World War was to intervene before nationalisation of the British coal industry provided the opportunity for coordinated action by both mining engineers and doctors to deal with the problem, as recounted in the next chapter.

Chapter 7

What does coal do to miners' lungs? The Cardiff studies

During my time in West Virginia I had been working on methods of early diagnosis of disease in coal miners' lungs by physiological tests and by a standardised method of reading changes on X-ray films. But after nearly two years I decided to return to follow my ambition to work as a consultant in the NHS, and the job I obtained was in Cardiff, South Wales, where I was appointed in 1971 as a chest physician, initially at Sully Hospital and later also at Llandough Hospital. These hospitals were situated on the edge of the Vale of Glamorgan, and looked south over the Bristol Channel. The fertile vale was situated on fossiliferous limestone at the southern end of what 15,000 years before had been the vast northern ice cap. To the north were the Rhondda valleys, sculpted by that ice, where coal had been mined since the early nineteenth century, to be brought by railway for export from the docks in Cardiff and Barry. This South Wales coalfield, the most important and productive in Britain, had suffered during the Depression of the 1930s but revived somewhat after nationalisation in 1946. Now, in the 1970s it was at the start of its terminal decline.

Sully hospital, inevitably now sold to developers and converted into luxury flats, was an Art Deco building with spacious grounds. It had served as a sanatorium, and over 150 TB patients were still treated there each year. The old days that required the patients to spend two or more years in hospital were being superseded by the use of effective drugs, which at that time required the patient to spend only one month in hospital, although 18 months' carefully supervised continuing treatment at home was needed for a cure. The large lawns on which the patients would gradually increase their activity on recovery were now recreational fields for the staff in which the medical superintendent, Dr Bill Foreman, with the help of his staff, had built an outdoor swimming pool. Bill Foreman was a New Zealander who had had a trial for the All Blacks rugby side before enlisting in the Army as a medical officer in the Second World War. He had been captured in Crete and spent the war looking after sick prisoners, for which he was appointed MBE, learning German and Russian in the process. He then trained in London as a chest physician before coming to Cardiff.

When I arrived, Sully was the regional cardio-thoracic centre, and the grounds also provided pasture for a few sheep looked after by the only NHS shepherd! They were there for surgeons to perfect their heart surgery and had their own operating theatre. In addition to looking after patients with TB, we also treated the range of lung diseases and particularly lung cancer. With the decline in TB admissions, at first we had ample beds, facilities and staff, and the hospital grounds also had a cottage in which relatives of heart surgery patients could stay. It was a brief period when all seemed well with the NHS.

Shortly after arriving, my colleagues and I acquired a ward and out-patient clinic with a lung function laboratory at Llandough Hospital, and we were able to play a full part in treating acutely ill chest patients, taking a special interest in asthma patients, who were then not very well managed. Llandough Hospital was also the site of the Medical Research Council's Pneumoconiosis Research Unit (PRU) and had a ward directly responsible for the care of miners with lung disease. By chance I had arrived at the very place where almost all the twentieth-century research into coal miners' lung disease had taken place and where the unit on which the West Virginia one had been modelled was situated. To understand why the PRU was here, in South Wales, it is necessary to go back to the 1930s.

The early Cardiff investigations of coal miners

In the early twentieth century there was, as noted in the last chapter, a very high mortality from TB in South Wales, and a charitable association, the King Edward VII Welsh National Memorial Association (WNMA), had been established in 1912 to care for these patients. The moving spirit and major donor behind it was the philanthropist and founder of the League of Nations, the first Lord Davies of Llandinam, who had been a Liberal Member of Parliament; he had also founded a professorial chair in tuberculosis at the Medical School in Cardiff. His father had made his fortune as the principal coal mine owner in the Rhondda Valley during the Industrial Revolution. Following the introduction of X-rays to medicine in the 1920s, the doctors of the WNMA started to find many miners who had abnormalities on their chest films but seemed not to have TB, and this attracted the attention of the first Professor of Tuberculosis in the medical school, Stevenson Lyle Cummins (1873–1949), mentioned in the last chapter.

Cummins would have been likely to have read the standard textbook of medicine of the time, *The Principles and Practice of Medicine*, by Sir William Osler (1849–1919). Osler had been professor of medicine in Montreal, then in Baltimore, finally becoming Regius Professor of Medicine in Oxford in 1905, a post he held until his death. In 1918 he published the eighth edition, in which he made clear his view that coal caused a disease of the lungs, which could on occasion progress from patchy nodules to large masses of fibrous tissue that could break down from loss of its blood supply and cavitate.[1] This was the cause of the expectoration of black sputum that had been noted by doctors previously as indicative of coal miners' lung. He also pointed out that disability came late in this process, and that miners seemed to be relatively resistant to tuberculosis. When breathlessness occurred he stated it was due to concomitant emphysema. He had expressed these views consistently in his book back to the first edition in 1892, so had obviously been aware of the early Scottish reports of the disease discussed in the last chapter. In general, textbooks such as this express the accepted knowledge of the time, based partly on personal practical experience supplemented by reading the relevant literature. Osler quoted some patients he had looked after, but his personal experience of coalminers' diseases was probably very limited. As time passed, Osler's views have been proved correct, but ironically at the time most who read them would probably have regarded them as out of date.

Lyle Cummins had qualified in medicine in Ireland and joined the Royal Army Medical Corps. He had almost continuous experience of war under Lord Kitchener, serving first in Egypt and the Sudan and then throughout the First World War, becoming a colonel and Assistant Director General of Army Medical Services. In 1919 he became Professor of Tuberculosis in Cardiff, where he set about investigating coal workers' lung diseases. He published the classical descriptions of the pathology of coal worker's pneumoconiosis in 1930 and 1936.[2] He stressed the exposure to coal dust without quartz that occurred in day-to-day work underground and produced the first classification of the pneumoconioses, based on both pathological and X-ray appearances:

Type 1 – dust retention and diffuse symmetrical nodular fibrosis, as in silicosis;

Type 2 – as above, with the addition of irregular diffuse shadows, as in a mixed exposure to coal and silica, silico-anthracosis;

Type 3 – partial mottling with only dust retention – anthracosis; and

Type 4 – extensive unsymmetrical local accumulations of dust like cricket balls – as found in heavily dust-exposed workers, usually with complicating TB.

Anyone today examining the lungs of coal miners would be able to recognise these descriptions (Figs 7.1 and 7.2). Cummins' attempt to differentiate the X-ray appearances seems to have been made in order to fit with current concepts of its relationship to tuberculosis, which was not only rife in South Wales but also commonly co-existed with pneumoconiosis. Radiologically at the time, it would have been difficult to differentiate between the two diseases. Moreover, he was not (in those early days of radiology) in a position to describe the sequential changes of the disease or diseases that would occur over time. Thus in 1927 he remained sceptical of the role of coal itself in causing clinical disability. His views were also shortly to change, and in this he was much influenced by the seminal observations of E.L. Collis, whose own scepticism we met in the previous chapter, and J.C. Gilchrist.

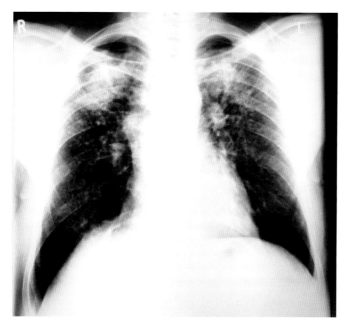

Figure 7.1 Chest radiograph of a coal miner, showing 'cricket ball' appearances in both lungs. These were later described as progressive massive fibrosis (PMF).

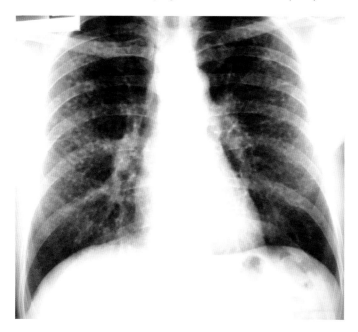

Figure 7.2 Simple coal workers' pneumoconiosis, showing 'snowstorm' appearance in the lungs. This appearance is generally not associated with disablement but is associated with a high risk of progression to PMF.

Edgar Collis was Professor of Preventive Medicine in Cardiff. He had been a notable athlete as a young man, captaining the Cambridge University athletics team and also swimming for the university. He had been the second doctor in Britain (after Sir Thomas Legge) to be appointed a medical inspector of factories and had been heavily involved in investigating the dangerous industries that poisoned their workers with such metals as mercury and lead, and in pushing for implementation of regulations to prevent the consequential illnesses. He became the UK's leading authority on bronchitis and dust diseases, and one who was not afraid to challenge Haldane's views on the lack of toxicity of coal. In 1923 he had published an article in which he pointed out that chronic bronchitis was likely to be a result of persistent irritation rather than infection or changes in temperature, pointing to exposures to gases and fumes in industry (and incidentally paving the way towards future studies of air pollution).[3] In 1928, with James Gilchrist, a tuberculosis physician of the WNMA, he published a study of the X-ray appearances of men employed as coal trimmers in Cardiff docks.[4] These men were employed in the holds of ships, shovelling the washed coal as it cascaded down in order to ensure that it was evenly distributed, and were thus exposed very heavily to dust that was pure coal with virtually no silica content. In spite of this they developed X-ray signs identical to those of underground workers. Moreover, several were disabled and none had bacteriological evidence of TB. The summary of their paper is worth quoting verbatim:

> Statistics are first quoted to show that coal trimmers do not suffer any excess mortality from phthisis, but that they do experience unusually high mortalities from both pneumonia and bronchitis. Next, as the result of clinical and radiographic examination, the case is maintained that these men exhibit a condition closely similar to that found among men exposed to silica dust. Hence the authors hold that other dusts than silica can so alter the lungs that they show 'snow storm' shadows when X-rayed, while the men are predisposed to respiratory diseases, but have no special tendency to succumb to tuberculosis. This conclusion is important since a tendency exists, as compensation for silicosis is being extended, to base a diagnosis of silicosis upon X-ray findings.

At this time Collis's colleague, Cummins, was inclined to attribute the lesions in coal workers to blockage of the draining lymphatic channels in the lung and damage to the associated lymph nodes, an observation that was almost certainly correct.[5] Perhaps his most important contribution, however, was his observation on the presence of emphysema in the lungs of coal miners, to which I shall return in the next chapter.

Cummins' work was continued by the Professor of Pathology in Cardiff, Jethro Gough (1903–79), whose most important contributions supported the presence of two types of pneumoconiosis in coal trimmers and miners, simple pneumoconiosis with only discrete small nodules, and a 'complicated' type in which there were large masses. These are analogous to Cummins' types 2 and 3, now called simple pneumoconiosis (Fig. 7.3) and type 4, now called progressive massive fibrosis or PMF (Fig. 7.4). Cummins's type 1 would now be recognised as silicosis. Gough used a method of displaying whole lung slices on a paper background, a technique invented by his technician J.E. Wentworth.[6] Gough was somewhat inflexible and a dominant figure in the Cardiff medical school. He continued to believe that the PMF lesions were a consequence of concomitant TB infection and he came into dispute with the views of some of his colleagues on this. However, as a native Welsh speaker who had been

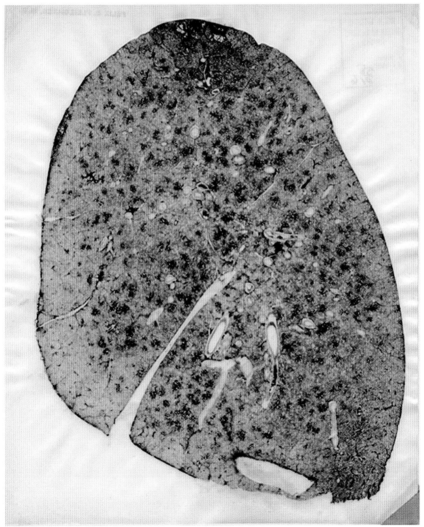

Figure 7.3 Gough-Wentworth section of lung showing black coal deposits (macules) of early simple coal workers' pneumoconiosis. (Courtesy of Prof. Ken Donaldson.)

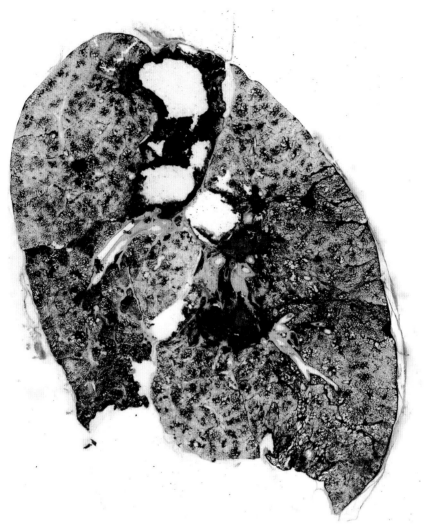

Figure 7.4 Gough-Wentworth section of lung showing black deposition of coal in central lymph nodes and PMF in the upper part. The PMF has cavitated and in life the miner would have expectorated black material. (Courtesy of Prof. Ken Donaldson.)

born in the South Wales valleys, he developed good relations with the mine workers' union and was a strong advocate for the award of industrial injuries benefits to coal miners for pneumoconiosis in the absence of silicosis.

The Medical Research Council: recognition of coal workers' pneumoconiosis

The dawning realisation among doctors that coal dust exposure *per se* could cause severe and sometimes disabling lung changes supported the long-held belief of the miners and their Trade Union representatives, who had been calling for coal worker's lung to be made compensable under the Industrial Injuries Regulations alongside silicosis. There was pressure on government to do something about the likely increasing burden of having to pay compensation to large numbers of mine workers, especially as the signs were that a war with Nazi Germany was looming. The mining industry was clearly going to be vital to any war effort, so the UK Medical Research Council (MRC) set up an Industrial Pulmonary Diseases Committee chaired by someone whose interests lay in diseases of the intestines, pretty well the only organ in the body that is not affected by industrial disease! Initially this committee was sceptical of the existence of a separate disease, being rather negative about it at an international mining conference in South Africa in 1930.[7] However, following the work of Cummins and Collis and in recognition of the size of the problem, in 1937 it was asked by the government to investigate coalmining urgently. In 1942 Philip D'Arcy Hart (1900–2006) was asked to carry out the study.[8]

Hart was a physician who had done something remarkable for the time: in 1937 he had moved from his secure and potentially lucrative consultant post in London to work in research for the MRC, and was immediately involved in this investigation with another tuberculosis doctor from the WNMA, Edward A. Aslett, and a number of specialists in physics, pathology and minerology. Following this, Hart became director of the MRC tuberculosis research unit until 1964, and then continued working in tuberculosis research for the MRC and on research grants until his death at the age of 106. In his career he made seminal contributions in the area of controlled trials of treatment, especially of TB, but his best-known work was in his early studies of coal miners with Aslett and others, published in a series of three MRC reports between 1942 and 1945.[9, 10]

The first MRC investigation was a preliminary survey of miners at Ammanford pit in the west of the South Wales coalfield, subsequently extended to 13 other pits, including the high-rank anthracite in the west and lower rank bituminous and steam coals in the east. They confirmed the work of Collis and Gilchrist and also showed that pneumoconiosis was most prevalent among the anthracite miners and, particularly importantly, that the higher the coal rank in terms of carbon content and thus combustibility, the higher the risk to miners of developing the disease. They were able to describe progressive changes on the X-ray films of miners, but were handicapped by the poor sensitivity of their apparatus when describing the earliest changes. And by testing the sputum of their subjects for mycobacteria, they were able to exclude TB as responsible for pneumoconiosis. Alongside this study, the MRC also carried out toxicological studies of coal dust by inhalation studies in rats, studies of the lungs of miners, and of dust levels in mines. Together, all of these studies contributed to the understanding that coal mine dusts differed in their harmfulness in relation to coal rank, but also that overall, the higher the dust levels in mines, the greater was the risk.

Taking account of the fact that coal dust exposure occurred in coal trimmers in ships' holds and in workers on the surface screening coal to remove impurities, coal *workers'* (rather than only coal *miners'*) pneumoconiosis (CWP) was recognised as a scheduled disease for compensation in 1934. This recognition that coal workers' pneumoconiosis was a disease in its own right, separate from silicosis, welcome as it was to the workers, also immediately revealed other problems that required solution. It had been more than an academic exercise; disabled human beings were at the heart of it.

Two problems were particularly pressing: first, certifications of disability under the regulations started rising at an alarming rate, from around 200 a year in 1935 to over 5000 in 1945. Given that certification led to loss of job as further dust exposure was thought inadvisable, this produced a crisis of unemployment in mining areas, made even worse by the economic depression of the 1930s and only relieved temporarily during the 1939–45 war when conscription of non-miners into the armed services made surface jobs, such as in factories, available for those displaced by the disease but still physically able. However, the sheer size of the problem thus revealed meant that serious preventive action had to be taken.

Secondly, and confusingly, it became apparent that many men with radiographic pneumoconiosis had little or no actual disability, while some men with no pneumoconiosis were in fact disabled. These issues were of international importance in view of the role coal then played in the economy of the developed and now regenerating post-war world; similar reports of so-called 'silicosis' in coalminers had also been published in Europe, Australia, the USA and Russia during the 1930s.[11]

The formal recognition of coal workers' pneumoconiosis as an occupational disease separate from silicosis was a long-delayed and satisfactory conclusion to an argument that had lasted over 100 years. This argument was characteristic of an insidious and persistent problem in science, and indeed in many other fields: a battle between conservative authority and dangerously original thinking. Theories are proposed and their originators may take them in their arms and protect them from attack rather than subjecting them to the destructive testing that science demands. Often, expert committees comprise the more conservative element and, while this may prevent some errors, all too often advances in knowledge are delayed.

In the case of coal mining, a few individuals with intimate knowledge of the working men and their conditions eventually changed scientific understanding. The first objective of the researchers was to prevent the disease from occurring rather than to enable compensation of those who were already suffering. The outcome of the early research was the inevitable plea of the scientist: more research was required, and this implied more funding. While in a privately owned industry this was not very welcome to the employers, in the case of coal miners' lung diseases, it was fully justified on public health as well as economic and humanitarian grounds.

The MRC Pneumoconiosis Research Unit

In 1945 the MRC established a special unit, the Pneumoconiosis Research Unit (PRU) to continue these investigations under Charles Fletcher (1911–95).[12] His selection for the post was not untypical of the time, a method that indeed persisted into the 1960s. The head of the MRC, Sir Edward Mellanby, had ordered a search for a suitable doctor, but this was unsuccessful, perhaps because a move from clinical practice, which was essentially private before the NHS, was financially unappealing. On learning this, he told the chairman of the search committee that the son of his predecessor as the head of MRC was just the man, even though he knew nothing about pneumoconiosis! At first sight he must have been thought by his colleagues an unlikely person to be invited to work in South Wales in mining communities – tall, very good-looking, highly intelligent, he had been educated at Eton and Trinity College Cambridge and had rowed (successfully) for the University in the 1933 boat race against Oxford. A remunerative career in consultant medical practice must have beckoned. However, he proved an inspired choice, being in political sympathy with the miners and having a natural ability to get on with people.[13] Later he put his communication skills to good use in the early days of television by presenting a documentary series on medicine, *Your Life in their Hands.*

It must be remembered that these were difficult times in Britain. The country had been impoverished by war and was reliant on loans from the USA for reconstruction. Food was scarce and rationed – indeed the staff of the Unit continued to forego butter and margarine at their breaks for at least a decade after rationing ended in 1952! Having learnt a hard lesson from Hitler on the consequences of ultra-right wing politics and suffered through the depression of the 1930s and the subsequent war, the British people in 1945 voted overwhelmingly for change to a Labour government under Clement Attlee. This government instituted the Welfare State and a programme of nationalisation, including the mining industry, which in 1946 became the National Coal Board. In 1948, influenced by the Beveridge Report,[14] the government founded the National Health Service, with Beveridge's aim *to create comprehensive health and rehabilitation services for prevention and cure of disease.* The original staff of the PRU by and large reflected the radical ethos of the times.

Fletcher was joined by some very distinguished scientists, the most notable of whom was Archie Cochrane (1909–88). Born in Scotland, he had won a scholarship to King's College, Cambridge, but his anti-fascist sympathies led him to serve in a rather chaotic ambulance brigade during the Spanish Civil War before qualifying in medicine in 1938.[15] As a medical officer in the Second World War, like his friend Bill Foreman he was captured by the German Army in the retreat from Crete and spent the war trying with no facilities to look after his fellow prisoners, including Russians, who were treated particularly badly, ravaged by TB and other infectious diseases and suffering starvation oedema. His autobiography gives a graphic account of these times and how his natural desire to find scientific answers to problems led him to do his first clinical controlled trial in awful circumstances. His views on evidence-based medicine were much influenced by his experiences. After his time in the MRC Unit he became the third David Davies Professor of Tuberculosis and made highly significant contributions to improving the practice of medicine by attention to its evidence base and the use of the randomised controlled trials he had pioneered. Regrettably, he clashed with Jethro Gough, by then the dominant personality in the Medical School, over their contrasting views on the role TB played in pneumoconiosis. On his retirement he recommended that his chair be discontinued, as TB was very much on the wane. In this he missed an opportunity to keep Cardiff in a leading place in research in the changing patterns of respiratory disease.

Many of the original members of the PRU were distinguished graduates of Cambridge or Oxford universities, reflecting a characteristic of the times that made it seem a rather elitist and English body. This may explain some of the tensions that were to develop between it and the Welsh Medical School. Two original members were John Gilson (1912–89) and Philip Hugh-Jones (1917–2010), physicians who had a particular interest in the lungs and how they work. Gilson had researched oxygen and gas masks for the RAF during the war and was a practical man who made his own experimental equipment when necessary. When Fletcher moved back to London, Gilson succeeded him as Director. Hugh-Jones (1917–2010) was the illegitimate son of a Liberal MP, his mother being associated with the Bloomsbury group. Like Cochrane, he had been a scholar at King's College Cambridge, and after his time at the MRC became a distinguished chest physician and physiologist in London. Together they pioneered new ways of measuring lung function, notably the measurement of total lung volume, before Hugh-Jones went to the West Indies and then back to London to pursue twin careers as medical researcher and anthropologist. Two physicists also played important roles in the early days of the PRU. Margery McDermott invented the first spirometer not requiring a water gas meter, making it suitable for large community surveys, and Vernon Timbrell devised methods for exposing rats experimentally to measured amounts of dust in order to study the pathological reactions to different types of dust.

A special mention should be made of Alice Stewart (1906–2002), who was briefly a member of the PRU and published an important study in 1948 that showed the fatal consequences of having advanced pneumoconiosis; symptoms and x-ray abnormalities progressed even after dust exposure ceased.[16] Alice Stewart was a pioneering female doctor at a time when women were very rare in British medicine.

Indeed, in 1956 when I entered medical school, fewer than 10% of the students were female and there was an even smaller proportion among our teachers. Now female students are in the majority. Stewart had already done important work during the war on poisonings in munitions factories, and was to go on to become the world-leading expert on effects of radiation, for example of maternal radiation during pregnancy in increasing risks of cancer among their children. She was under-appreciated by the British medical establishment but continued working on grants from the USA, where she was better appreciated in the era when nuclear war threatened, the so-called Cold War.

Other early members of the PRU included the statistician Peter Oldham who devised improved ways of analysing data on dust exposure of miners, in those days when there was little assistance other than slide rules, logarithms and simple mechanical calculators, and Martin Wright (1912–2001), originally a pathologist, but who became best known as an inventor of, among other things, the peak flow meter, the breathalyser, and an alarm to prevent cot deaths.

Achievements of the PRU research

All these medical scientists went on to make major contributions to health care generally, but with respect to the understanding and prevention of coal workers' pneumoconiosis the most important studies related to the design and performance of epidemiological studies led by Cochrane. These led to a much better understanding of how to investigate the way the lungs, and by implication other organs, may be damaged by environmental factors:

- The recognition that bias in collecting data can play an important part in influencing the results of studies;
- The importance of obtaining a very high response rate from participants in population studies;
- The need to use standardised methods of measurement and to test their repeatability; and
- The development of instruments to measure lung function and dust concentrations

The studies of miners in the South Wales valleys, led by Cochrane, were an exemplar of how to perform epidemiology: the researchers went to endless trouble to ensure full participation. One very important outcome of the research was the description of the way in which coal workers' pneumoconiosis presents and progresses on the chest radiograph. Their use of improved radiological techniques allowed them to define clearly the stages of coal workers' pneumoconiosis. For this they selected X-ray films of miners showing increasing density of the 'snowstorm' appearance of small opacities and different sizes of the 'cricket ball' lesions. The former films, snowstorms, were divided into categories 1, 2 and 3 by increasing profusion of opacities (see Fig. 7.2). The cricket balls were categorised as Stage A, B or C with increasing area on the film. This was fundamental to studies of risk in relation to exposure to dust, allowing for comparisons between observers and between subjects, and thus promoting more consistent and reproducible reporting of results.

Fletcher gave the name Progressive Massive Fibrosis (PMF) to the most severe form, the 'cricket balls' of Collis and Gilchrist (see Figs 7.1 and 7.4). The earlier 'snowstorm' stage in which there are only small spots or nodules on the X-ray film was referred to as simple CWP.[17, 18] In due course, in international collaboration, a series of standard films was produced by the International Labour Office (ILO) along with detailed instructions on reading them for epidemiological purposes.[19] These remain the standards by means of which all films of individuals with suspected pneumoconiosis are classified, and allow the reader to categorise the films as showing increasing densities of spots as stages 1 to 3. Use of this technique allowed the PRU to show that simple CWP was not associated with dysfunction or with premature mortality. It also proved possible to show that, nevertheless, the main risk factor for PMF was the presence of simple CWP.[20] For the first time it had become possible to quantify pneumoconiosis in terms of the severity of X-ray change, leading the way to relating risk to quantified dust exposure.

A fundamental requirement in epidemiology is for all measurements to be reliable and reproducible. A standard symptom questionnaire for studying lung disease was developed by the PRU and was widely used subsequently in population studies of bronchitis and emphysema, allowing the results in different places to be compared. New methods of measurement of lung function, especially the one that relates most closely to breathlessness, the FEV_1, were also applied to epidemiological studies, and rules were devised to improve their reproducibility. It can reasonably be claimed that modern respiratory epidemiology started in Cardiff.

Fletcher left in 1952, to make further seminal contributions to the understanding of the epidemiology of chronic lung disease in the general population and to lead the campaign against the cigarette companies who were responsible for so much illness and death. The PRU continued under Gilson, who had a rather hands-off approach. The epidemiology was led by Cochrane who, on being appointed to the David Davies chair of TB in the Medical School, established a separate MRC Epidemiology Unit. From this time the multi-disciplinary teamwork that had characterised the PRU in Fletcher's time disappeared and each individual senior member of staff pursued his or her own interests. The clinical and physiological studies were led by John Cotes and Colin McKerrow,[21] and they commenced a series of important studies on oxygen therapy and rehabilitation of miners and others with chronic lung disease, and of lung function in different populations, which continued until Cotes left in 1980 to continue his research in Newcastle University.

Cochrane and the epidemiology of CWP

From the point of view of CWP, the most important work was that of Cochrane and his team in the 1950s. In order to understand the relevance of this and why it led both to clarification of the main issues and also to further medical disputes, it is necessary to understand a bit of the difference between epidemiological and clinical approaches to medicine and Cochrane's role in this. I have introduced this subject in chapter 5. In the early twentieth century there was a rather dominant hierarchy of physicians and surgeons in medicine centred on the Royal Colleges in London and Edinburgh. Clinical practice was based very much on experience, that is, what individual doctors had found to be effective. Unfortunately, experience can be and often is fallacious. Cochrane was a natural sceptic and questioned everything – something that can easily lead to conflict with one's colleagues. He was also a lateral thinker, readily coming up with ideas that others had not conceived, and this also easily creates opponents among colleagues whose ideas may have become fixed. And, as I have shown, the diseases of miners had already generated much dispute.

The most important issue in coal workers' pneumoconiosis at the time was PMF. The early studies had shown that what became known as simple pneumoconiosis of coal miners, even with quite obvious small spots scattered throughout the lungs on X-ray, was often consistent with full fitness and normal lung function. But when the X-ray showed large masses, often surrounded by emphysema, the miner became progressively more disabled and often died of lung failure. Gough was convinced by his pathological studies that the cause of this was tuberculous infection that prevented the lungs getting rid of the dust by obstructing the lymphatic channels.[22] Cochrane now had the possibility of seeing if this was in fact the case, and devised an ambitious study to do so. He decided to X-ray all adults (not confined to miners) in two South Wales valleys, each with four pits and comparable social conditions and numbers of miners, in order to detect cases of both pneumoconiosis and TB. By sheer determination, hard work and with the support of the National Union of Mineworkers, he achieved this and found a frightening rate of TB. His plan was to remove those with active TB in one valley for treatment (at that time in a sanatorium for bed rest, fresh air and the newly discovered streptomycin injections), leaving those in the other valley to have the standard management provided by the new NHS. The intention was then to re-survey all participants and to see if the rate of PMF was greater in the treated or untreated valley.

As it turned out these surveys, though failing in their primary objective, did produce very important findings; development of PMF was shown to be a significant risk in men with the more advanced category 2 (out of 3) of simple pneumoconiosis and progressed more rapidly in younger men than in older, but was not influenced by smoking. Although the survey was not able to answer convincingly whether TB played any role in the causation of PMF,[23] this understanding of the natural history of pneumoconiosis led to what was probably the most important practical result of the research, the ability to reduce risks of disabling disease by X-ray surveillance of miners. Cochrane argued that regular chest radiographs, read by specially trained doctors, could detect the early signs of pneumoconiosis before disability occurred and that redeployment of such workers could prevent further development to the disabling stage of PMF.[24]

Cochrane's studies were less successful with respect to the effects of dust exposure on the function of the lung. This probably stemmed from the fact that, at the time, epidemiology was regarded by the MRC as a rather less interesting scientific activity than physiology; Cochrane thought the other early senior members of the PRU were too interested in complex laboratory physiological tests, which were insufficiently portable to be used in population surveys. Nevertheless, as mentioned above, they were to develop the reliable portable spirometer a bit later, and had already carefully investigated the reliability and reproducibility of measurements of lung function.[25] This led to the adoption of the FEV_1 as the standard test for use in epidemiology. The PRU did also make another very significant advance of relevance to epidemiology. Peter Oldham and a physicist, Bill Roach, devised a method for estimating the dust exposures of miners, using a combination of measurement of the concentrations in mine air, using an instrument called a thermal precipitator, with a statistical means of estimating the exposure of groups of miners over shifts. Using this method, Cochrane's group made the first estimates of the risk of developing PMF in relation to dust exposure. However, their intention to study the relationships between dust exposure and risks of CWP in 20 pits across the country was taken over by the National Coal Board in the early 1950s, as recounted in chapter 8.

After Cochrane obtained the Chair in Tuberculosis, influenced by the early trials of anti-TB treatment he turned his attention increasingly to randomised controlled trials and other important projects unrelated to coal, and achieved international fame from his promotion of evidence-based medicine, his name commemorated in the Cochrane collaboration that oversees large-scale trials of therapies in the UK. There is an interesting irony here: he is now regarded as the high priest of evidence-based medicine, an orthodoxy that is not to be questioned, yet he was himself the most sceptical of men. I think, were he alive, he would be the first to question the evidence base on which pronouncements on patient management are now made: in many cases it remains insecure.[26] Part of the reason for Cochrane's diversion from coal research was opposition from Gough who, as noted above, was a dominant figure in the Cardiff Medical School; but more important was the decision made in 1969 by the MRC to hand over research into coal miners' diseases to the National Coal Board itself. It reasonably held that the newly nationalised industry should be responsible for the research required to prevent the diseases it was responsible for causing.

The final phase of the PRU

The PRU received a boost when the pathologist Chris Wagner (1923–2000) arrived in 1962 from South Africa, where he and his colleagues had described the association of the fatal cancer of the lining of the lung, pleural mesothelioma, with exposure to blue asbestos. This had attracted worldwide interest since, although the tumour was then rare, it rapidly became apparent that many millions of people had been exposed to asbestos and could be at risk. Wagner was sociable and, in contrast to some of his colleagues, well able to work in a multi-disciplinary team. He shared with several of them experience of war, having interrupted his medical studies to fight in the South African army in North Africa and Italy, including at the battle of Monte Casino. He collaborated especially with physicist Vernon Timbrell, mineralogist Fred Pooley and statistician Geoffrey Berry, and this ensured a new lease of life for the PRU related to a different pneumoconiosis, asbestosis, and similar diseases.

However, Wagner also took an interest in coal workers' pneumoconiosis and collaborated with a young MRC trainee physician, Anne Cockcroft,[27] and an NHS pathologist, Roger Seal, in two important studies.[28] In the first, published in 1982, they examined the lungs of men who had died of heart attacks and made quantitative measurement of the amount of emphysema in them, showing that it was more common and more severe in miners than in non-miners.[29] In the second, in 1986, they investigated the role of the lymph nodes at the root of the lung (the hilar nodes) in PMF and showed that the more severe the inflammatory change in them, the greater the likelihood of PMF,[30] thus looking back to the earlier work of Cummins on the lymphatic vessels and their role in pneumoconiosis. If the lymph nodes are destroyed by inflammation, the route for removal of dust deposited in the alveoli is blocked. The dust will then accumulate in the lung tissue and, with the accompanying inflammation, cause the large destructive masses of PMF. As TB also causes destruction of lymph nodes, it seems likely that in the early days of Cummins and Gough it could well have played a role in some cases of PMF.

On Gilson's retiral in 1974, Prof. Peter Elmes, who had worked on the asbestos-related tumour, mesothelioma, in Belfast, became director, but by then the primary objectives of the Unit with respect to coal pneumoconiosis had largely been achieved and most of the senior figures were retiring or had moved elsewhere. The Unit continued with interests in asbestos and other causes of occupational disease until the MRC finally closed it down in 1984; its legacy was assured by its work on coal and asbestos and continued in its many ex-members of staff scattered around the world in academic medical research. Its history and achievements have been well summarised by John Cotes.[31]

In summary

By the 1980s, the work of the physicians and pathologists in Cardiff had done much to elucidate the effects of coal dust on the lungs and to point in the direction of prevention of the diseases. They had shown that pneumoconiosis of coal workers was essentially benign in its early, simple form, but that miners who developed this were at greater risk that it would then progress to PMF. They had shown that PMF tended to get worse even after dust exposure ceased. This led to advice that miners should be X-rayed periodically, and those with early changes should be excluded from dust exposure. They had suggested that PMF was caused by heavy dust exposure and had contributed to understanding how to measure such exposures underground. They had shown that emphysema and bronchitis were common in miners but had not yet convinced the medical profession, or indeed themselves, that this was not entirely due to smoking. What remained was to quantify the risks to miners, taking account of the different mining conditions across the UK, in order to assist in preventive action and to investigate whether the disability from emphysema was caused by coal dust exposure. There was also the question as to why such different rates of pneumoconiosis occurred in different mines; what was it, in different coal dusts, that was responsible? These tasks, together with the most important job of reducing the risks, now appropriately became the responsibility of the nationalised coal industry itself.

Chapter 8

Tying it all up: bronchitis, emphysema and pneumoconiosis

Chance had played a large part in the way my career put me close to coal. Early on I had spent a year in Stoke-on-Trent in the English Midlands as a medical registrar, a trainee physician. Stoke is the site of the famous pottery industry, established there by Josiah Wedgewood in the eighteenth century because of the abundance of local coal and clay, even though the latter had subsequently to be brought up from mines in Devon in southeast England. This was one place where doctors saw both coal miners and potters with pneumoconioses, the latter with silicosis from the use of crystalline silica in making and firing the china. It was there that I first saw patients with industrial disease. One such, a retired potter, had an unconnected condition, a stomach ulcer which had caused him to vomit blood. I asked him if he was taking aspirins, since at the time these were commonly used painkillers and were well known to cause stomach bleeding. He was, and when I asked him why he took them, he told me it was to ward off the dust.

This was in the days before it was known that aspirin has an effect in reducing the risks of heart attack, and I had not previously heard this local folklore, that aspirins were thought by potters to prevent silicosis. Subsequently it was shown that aspirin is protective not only against heart attack but also against bowel cancer, which later led me to wonder if indeed, by its anti-inflammatory action, it might be preventive against pneumoconiosis also.[1] Of course, we shall never know, as pneumoconiosis is prevented by controlling dust exposure in the workplace. It is strange that aspirin (acetyl salicylic acid, a simple chemical originally derived from the willow tree, *Salix ripens*) should have such effects on human beings. It is presumably of some benefit to the tree, perhaps in deterring parasites.

Later, as recorded in chapter 5, came the spell in West Virginia where I worked on coal miners' diseases with Keith Morgan and his colleague LeRoy Lapp. It was there in 1969 that I asked myself the question, does coal cause emphysema? At the time I was ignorant of much of the research that I have summarised in the previous chapters, but I quickly realised how contentious the issue was. The unit I was working in, a branch of the US Public Health Service, was called the Appalachian Laboratory for Occupational Respiratory Diseases (Alford). It had been established essentially to replicate the British work on understanding the lung diseases of coalminers, but also to determine their prevalence across the USA and thus assist in their prevention. To that end, it was involved in large scale surveys of miners using questionnaires, X-rays and lung function tests. My little bit involved mainly lung function tests that might be helpful in diagnosing emphysema which, at that time before computerised tomographic (CT scan) techniques were available, was difficult to diagnose at an early stage.

While this was going on, the growing awareness of pneumoconiosis among US miners and their Trade Unions was leading, in the American style, to a profitable business for doctors and lawyers in certifying miners as disabled so they could receive compensation. This became apparent to the doctors of Alford when miners said to be totally disabled proved to have quite normal lung function when tested in the laboratory. Keith Morgan, who was a Yorkshireman, a race known for their bluntness, was sceptical and said so, attracting enemies. He was one of several very bright British specialists in respiratory medicine who had migrated to the USA in the post-war period, when the NHS had just started and most senior posts had been filled by relatively young men (but in those days rarely women), making consultant posts scarce.

Keith Morgan had set up the Alford research programme some years before, and when I arrived it was very active. As a small part of this we did some early investigations that showed that non-smoking miners with normal standard breathing tests such as the FEV_1 did sometimes have more subtle detectable abnormalities (raised lung volumes and changes in the elasticity of the lungs) that to me suggested early emphysema.[2] We also spent some time using an X-ray method of measuring lung volumes to see if it could be used in large community surveys. Keith Morgan remained sceptical of the possibility that coal dust caused COPD, preferring the view that the airflow obstruction was in general due to smoking. Some years after I had left, he moved to a university chair in Canada. It was rumoured that he had felt seriously threatened by the antagonism of the union, the United Mine Workers of America, which had a robust reputation.

By this time, 1972, I was in Cardiff working as a chest physician. I had hoped to work part-time in the Pneumoconiosis Research Unit, and Colin McKerrow had agreed to my working with him to continue this interest in emphysema. Very sadly, some weeks later he was found to have a fatal cancer and this plan fell through. But I had already decided to write a book on occupational lung diseases, since these conditions were then rather neglected in medical teaching, and Keith Morgan agreed that we should do it together. While we were involved in this it became clear that we held opposing views on the likelihood of coal causing emphysema. Nevertheless, at the time it was an open question for a researcher, and I was in no position, as a very busy NHS consultant, to find the answer.[3] I switched my research interests to trying to discover the most appropriate way to treat severe asthma attacks, a condition we saw frequently. Then again came the hand of chance.

In the summer of 1977 I received a phone call from the chief medical officer of the National Coal Board (NCB) asking if I would be interested in a job as Director of the Institute of Occupational Medicine in Edinburgh. He explained that this institute had been established as a charity by the NCB to investigate miners' diseases and that the previous director had left for an academic post in Canada. The staff were employees of the NCB. I agreed to visit the institute and to make further inquiries before accepting. In particular I hoped to be able to continue to work part-time in the NHS, but this time in Scotland, as a chest physician.

The National Coal Board's role in prevention of miners' diseases

The National Coal Board had been established in 1946, two years before health care in Britain was also nationalised under the Labour government of Clement Attlee. On nationalisation, the NCB had acquired 958 pits, 55 coke ovens, 30 smokeless fuel plants and some 800,000 workers, 4% of the country's workforce. Among them was an unknown but probably very large number of men with lung disease who, rather than being employed by one of 800 different companies, were now all employed by the State. It was known that pneumoconiosis was widespread, though its prevalence differed greatly between area and pits. The work of the MRC had shown that the advanced form of pneumoconiosis, PMF, was disabling, usually progressive and often fatal, and that the risk of developing this was greatly increased in men who had category 2 simple pneumoconiosis. But it was also clear that many men without pneumoconiosis in the industry became disabled by breathing difficulties, so there were two big medical questions to be answered alongside the major preventive question, how to prevent the disease yet continue to produce coal? These questions were:

- How much and what kinds of dust cause pneumoconiosis?
- What dust levels in mines need to be maintained if men are to be prevented from becoming disabled?

At the time of nationalisation the NCB had set up a medical service with two main purposes: to deal with accidents and injuries in the workplace (remember, this was pre-NHS) and to assist in solving the pneumoconiosis problem. The medical service was directed from 1951 by the Chief Medical Officer, Dr John Rogan (1913–89), and included a radiological service with mobile units distributed across the UK from Scotland in the north to Kent in the South, and medical centres at all the pits with first aid facilities and access to doctors and nurses. John Rogan had qualified in medicine in Edinburgh, served in the army Medical Corps in the war and, after working in industrial medicine in Glasgow, had a spell as head of the

MRC's environmental disease section. His military background served him well in addressing the pneumoconiosis problem.

Clearly the first priority was to detect and deal with cases of pneumoconiosis. A nation-wide surveillance programme was initiated whereby all miners were to receive a chest radiograph every five years. The doctors of the radiological service were trained in reading the films using the new standard procedure pioneered in Cardiff and soon to be accepted as an international standard under the International Labour Organisation. By the 1950s there was sufficient evidence to suggest that simple pneumoconiosis was rarely related to impairment of function, but that to have it entailed a significantly increased risk of progression to PMF, the serious stage of the disease. Previous legislation required all men with X-ray changes to be removed from work, but in view of the large numbers and their difficulty in finding alternative employment at the time, this was modified. Thus, a decision was taken that, from 1948, men who only had early signs of pneumoconiosis on X-ray (what were defined at category 1) could continue to work at the coal face but men with category 2 or more had to be offered alternative work. These men were eligible by then to claim disability benefits and were permitted to work in 'approved dust conditions', that is, away from the coal face underground or on the surface, where lower dust levels prevailed.

This important pragmatic decision to allow men with early changes of pneumo-coniosis to continue at work was possible because at that time it was recognised that early CWP on X-ray had very different implications from those of early sili-cosis. In spite of the fact that the appearances were very similar, with reduced coal dust exposure CWP would probably not progress, whereas even if silica exposure ceased, silicosis was likely to get worse. This broad generalisation turned out not to be completely true.

Alongside this medical management, the NCB instituted a programme of dust control and dust monitoring in all pits and two important programmes of research: to understand the engineering aspects of dust generation and control, and to under-stand miners' lung diseases. From the point of view of future medical research, the most important advance came not from the medical researchers but from physicists: how to measure dust in the air.

Measuring dangerous dust

The first thing a scientist wants to do when presented with a problem is to take some measurements. The problem, in essence, was to find out how much dust in coal mines was required to cause pneumoconiosis, and thus what levels of dust would be allowable to significantly reduce the risk of the disease occurring and yet permit coal to be mined. In the context of the newly nationalised industry this process required agreement from three parties: the NCB management, the trade unions (in this case principally the National Union of Mineworkers), and the Government regulator, the Mines Inspectorate. Representatives of these parties, plus the medical service and the researchers, formed a National Dust Prevention Committee to oversee the research into both medical and engineering aspects. This supervisory collaboration ensured that the research results were open and that ensuing action was agreed by all parties; it formed in its day a model for occupational medical research.

While it may seem fairly obvious that to measure dust in the air one simply has to catch and weigh the amount in a given volume, there are hidden complexities. What was known by the 1940s was that pneumoconiosis was caused by the dust that reached the furthest parts of the lung, the bronchioles and alveoli, where the damage had been observed by pathologists. In a cloud of dust, how do you measure the particles that can get to those tiny structures and exclude those that cannot? Physicists had shown that only particles of a certain size range, below about 7μm (micrometres) in diameter, are likely to reach the alveoli and be deposited there. This fraction of the total dust cloud became known as the respirable dust fraction. Up until this time measurement had depended upon time-consuming microscopic examination of particles,[4] a technique that had produced very variable results.

An important breakthrough came when Henry Walton and Robert Hamilton, two physicists working for the NCB at its Mining Research Establishment, invented a respirable dust sampler, the MRE 113a. This instrument sampled air that was sucked at a pre-determined flow rate through the gaps between a series of horizontal metal strips, an elutriator, and thence through a filter that caught the residual dust (Fig. 8.1). This was designed so that the larger particles fell by gravity

Figure 8.1 The MRE 113A respirable dust sampler. The elutriator is on the top and the entry to the left. Dust is sucked through this by a pump through a filter in the curved part to the right.

into the elutriator, in the same way that they fall from inhaled air onto the walls of the bronchial tubes, leaving only particles below about 7μm to collect on the filter. This instrument allowed the respirable dust to be weighed directly; if the sampler was close to a man at work underground the weight of dust recorded over his shift reflected the amount of dust he would have been likely to inhale that could reach the alveoli of his lungs. Thus, with an appropriate research strategy it would be possible to measure the exposures of many miners and to relate these measurements to outcomes such as symptoms, X-ray changes and lung function.

Measuring effects of dust on health:
the pneumoconiosis field research

The NCB's director of research from 1950 to 1963 was one of the best-known scientists of his era, Jacob Bronowski (1908–74). He had been born in what is now Poland and had come to England as a youth. After a scholarship to Cambridge he had obtained a PhD in mathematics. During the war he had worked on operational research for the RAF and then was called upon to investigate the effects of the atomic bombs in Japan. This turned him from military research to his pursuit of three major interests: biology, communication of science, and literature. His popular fame derived first from his appearances in the BBC's Brains Trust, where he showed his extraordinary range of knowledge and ability to explain the complex in simple terms. Later he presented the TV documentary, *The Ascent of Man*. His contributions to the NCB included the production of smokeless fuel, known at the time as 'Bronowski's briquettes', but also the institution of the research into pneumoconiosis.

The Pneumoconiosis Field Research (PFR) started in 1953. In the words of its original chief medical officer, Dr J.W.J. Fay:

> Pneumoconiosis is a condition which starts almost imperceptibly with the fixation of a small amount of dust, and as the period of exposure increases more and more dust is accumulated. So long as the condition remains as 'simple' pneumoconiosis, it is believed that it will not progress if the subject is removed from the dusty environment, and disability is absent or comparatively slight. The rarer but more serious form is known as 'complicated' pneumoconiosis,[5] or progressive massive fibrosis. In this form of the disease the patient's condition deteriorates even if he is removed from the dusty environment. Progressive massive fibrosis is thought to be caused by the superposition of an extraneous infection, probably tubercular, on a background of simple pneumoconiosis, usually in its more advanced stages. Hence if pneumoconiosis is halted at the earlier stages of the simple form the results are not serious. Both forms of the condition are recognisable by X-ray examination of the lungs.[6]

This was a fair statement of knowledge at the time, though the role of TB was beginning to be challenged. The objectives of the research, to determine how much and what types of coal caused pneumoconiosis and what dust levels should be attained in order to prevent miners from becoming disabled by the air they breathe, were both ambitious and clear. They were also remarkably ahead of their time, implying a requirement to measure both exposure and outcome in a large cohort of miners over a prolonged period, and to use these quantitative data to set protective health standards in the industry. Indeed, at the time of the enunciation of the objectives, it is doubtful if the computing power or statistical methods to handle and analyse the data were available anywhere.

Nevertheless, the researchers recruited some 50,000 coalminers from a representative 25 collieries across Britain, from Kent to Wales to Scotland, and started a programme of airborne coal dust measurement in all the pits. This was combined with five-yearly assessments of the miners by occupational and symptom questionnaires, lung function testing and chest radiography. The collieries were chosen to represent the range of mining and geological conditions in the country; thus started what was at the time the largest occupational epidemiological study ever carried out.

The PFR was unique not only in its size but also in the detailed estimates made of individual exposures of the 50,000 participants. Exposures are defined as the concentration of dust in the air multiplied by the duration of the individuals' presence in that environment, so it was necessary to know what all these men had been doing through their lives in the industry. Initially the measurements of dust concentrations in the air were made by an instrument called a thermal precipitator, which required counts by microscopy, but the MRE 113a sampler was introduced in 1965 and conversions made from the earlier measurements. A large team of technicians was employed at the pits to record the working practices of the miners and the length of time spent in different activities; the dust measurements were made on the basis of exposures of small groups of men. The measurements and assessments of exposure were made continuously for the duration of the research, ultimately 25 years. The medical data were obtained by mobile units of researchers every four years using symptom questionnaires (which included details on smoking), lung function by spirometry, and chest radiography. Initially a team of 115 investigators was employed in the research.

The foundation of the Institute of Occupational Medicine

The requirement to measure the environment as well as the outcome in humans involved a wide range of scientists in the research, notably physicists, statisticians, physicians, mining engineers and physiologists. With the support of the miners' trade unions, John Rogan persuaded the NCB's chairman, Lord Robens, to set up a special charitable research institute in Edinburgh to oversee the research and the many complex analyses envisaged. Thus it was that in 1969 this multi-disciplinary combination of scientists, engineers and technicians was grouped together to form the core of the original Institute of Occupational Medicine (IOM). IOM's foundation coincided with the receipt of a huge database, a landline link to the NCB's mainframe computer in England, and a headquarters building close to John Rogan's *alma mater*,[7] Edinburgh University, with satellite laboratories in the English and Welsh coalfields. The original senior members of staff, under John Rogan, were Henry Walton, Deputy Director and head of Dust Physics, who had co-invented the MRE 113a sampler, Michael Jacobsen, head of statistics, who had been a refugee from Nazi Germany and was a mathematician, and a chest physician and physiologist, David Muir, head of medical branch. Later, senior posts were filled by recruiting John Davis, a pathologist with special expertise in asbestos research, Jim Vincent to set up a physics branch with special interest in developing instruments for measuring the environment, Jim Dodgson to head occupational hygiene, Colin Soutar to head the medical group, and Tom Leamon to organise an ergonomics group.[8] All were to contribute importantly to the coal research as well as broadening the interests of the IOM.

The early research results

The initial objective was to analyse the data from the PFR with a view to determining the relationships between exposure to coal dust and risk of pneumoconiosis. Exposures may be expressed in various ways, but include both a duration and a concentration; in the analyses they were expressed as gram hours per cubic metre of air (gh/m^3), that is grams of dust in every cubic metre of air multiplied by the number of hours worked. The initial analysis of the first 20 pits to have completed three rounds of the survey was reported in the first important IOM paper, published in the prestigious science journal *Nature*.[9] It demonstrated a statistically significant relationship between the risks of developing simple pneumoconiosis and the cumulative exposure the men had had to respirable dust, such that a man exposed, for example, to $100gh/m^3$ over his working lifetime would have, on average, an approximately 10% risk of developing simple pneumoconiosis.

As expected, this risk differed considerably between different collieries and was later shown in an analysis of the NCB data across the whole British population of over 62,000 miners to relate significantly to the coal rank;[10] in this and all subsequent analyses the higher the rank, the greater the risk (Figs 8.2 and 8.3). Mining the best coal, in commercial terms, entails the highest risk of CWP. These statistical data represented the starting point for discussions between NCB management, the trade unions and the Government regulator on a level of dust in British mines that could be used as a standard. The maximum allowable

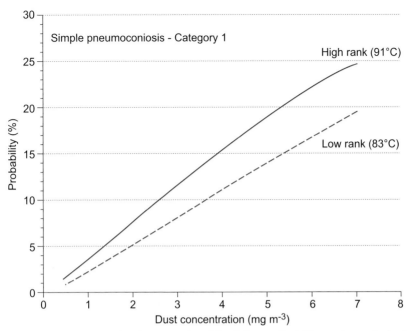

Figure 8.2 Relationship between the risk of the earliest X-ray signs of simple pneumoconiosis and average daily exposure to coal dust over 40 years underground. Note the difference in risk between high and low rank coal. Data from the Institute of Occupational Medicine.

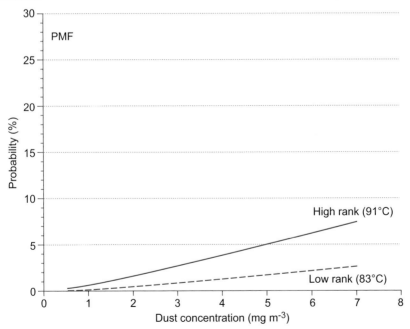

Figure 8.3 Relationship between the risk of progressive massive fibrosis and daily average exposure to dusts of different rank over 40 years. Data from the Institute of Occupational Medicine.

concentration was initially agreed, 8mg/m^3, later reduced to 7mg/m^3, as measured in the underground roadway (the outbye) leading away from the face. This ensured that it monitored the respirable dust generated by the cutting activities. The same data were then used by the USA as a basis for formulating their own coal mine dust standard. The implication of such a standard is that the risk to a man can be reduced in two ways concurrently: by reducing dust levels and by limiting the duration of exposure. After 15 years of the study, the primary research objective had been achieved.

A study by Jacobsen and his colleague Brian Miller of the mortality of these miners in relation to dust exposure had also confirmed the increased risks of death in relation to PMF but no excess mortality associated with the presence of simple pneumoconiosis, in line with Cochrane's findings at the MRC.[11] However, miners did have a significant excess death rate from chronic bronchitis compared to the general population; more comfortingly, they did not have an excess of deaths from lung cancer.

The year 1978 marked a watershed in the life of the IOM. Five successive huge nationwide surveys had been completed and data had been accumulated on the health and mortality experience of some 26,000 miners. In addition, it had proved possible, with appropriate permissions, to collect the lungs from autopsy of a large number of deceased miners who had previously participated in the surveys, and John Davis and his pathology group were studying these. But John Rogan had retired and his successor as director of IOM had resigned.

As I took up the post I sensed a feeling of uncertainty about its future, but it was apparent also that there were opportunities. The most important was to answer the questions that had plagued researchers hitherto about the inter-relationships of dust exposure, disability, and pneumoconiosis, using the rich data that had already been accumulated. The second was to plan for a future for the IOM and its staff on the assumption that the important questions on coal would soon be answered. But with these two opportunities came a major threat: the contraction of the coal industry in Britain and thus likelihood of reduction in funding and potential closure of the IOM.

Trouble and strife, and the decline of the UK coal industry

In retrospect it seems to have been seriously unwise of me to leave a secure job in the NHS to work for the NCB in 1978. Lord Robens as chairman had been faced with the need to reduce the size of the coal industry in response to cheaper foreign competition, attracting the enmity of the unions in return. His reputation had also been badly damaged by an insensitive response to the Aberfan disaster of 1964, and he had been succeeded in 1971 by Derek Ezra. Ezra had been a Cambridge history scholar who had enlisted in the army as a gunner in the war, but had rapidly been promoted to work in intelligence and had ended as a lieutenant colonel with an American Bronze Star. He joined the NCB as a marketing man and was again rapidly promoted, being appointed chairman in 1971.

The 1970s were an era of very high inflation and the Conservative Government under Edward Heath introduced a cap on public sector wages. Ezra's appointment had been followed in 1972 by a short miners' strike for higher pay, which he was able to settle. However, a second strike in 1974, against a background of an oil crisis and a government-imposed three-day week on electricity-consuming heavy industries, forced the defeat of the Government and a general election at which Harold Wilson's Labour Party was returned to power. The new government introduced a number of measures to protect the UK coal industry against cheaper foreign competition. Over this period Ezra had achieved a good personal relationship with Joe Gormley, who was the General Secretary of the NUM. Joe Gormley was an affable man who had worked his way up from the coal face in Lancashire through the union. He was less political but more intelligent than people thought, his motivation being the welfare of the miners, and this was reflected in their trust in him. In particular, he knew the way miners thought, and was well able to judge the likelihood and outcome of strike action.

The relationship between Ezra and Gormley was key to what happened next. In 1979 the Labour Government, now under James Callaghan and with no overall majority, fell and a Conservative government was elected under Margaret Thatcher. She had learned the lesson of the previous strike and was deeply suspicious of what she regarded as a too cosy relationship between the union and NCB's management, which she correctly saw as obstructive to the Conservative doctrine of privatisation of State assets. In fact, in Mrs Thatcher's early years as Prime Minister, Ezra had managed to persuade her to postpone proposed pit closures and to increase investment in the industry, averting another strike. However, Ezra and Gormley both retired in 1981, to be succeeded by Ian MacGregor (1912–98) and Arthur Scargill respectively. The scene was set for a major reduction in size of the UK industry, which incidentally had already contracted from 700,000 men in 1947 to 230,000 when Ezra retired.[12]

Ezra and Gormley were pragmatists, perhaps exemplified by the fact that both were appointed to the House of Lords on retiral, Ezra sitting as a Liberal Democrat. In contrast, MacGregor and Scargill had definite and contrasting political agendas which, though present in some of their senior colleagues previously, had been kept under wraps by their predecessors' effective management of industrial relations. From 1979, the year after my appointment to IOM, the writing on the wall for the UK coal industry was apparent. Thatcher planned like a general for the next strike, ensuring coal stocks were built up in power stations and neutralising the power of the nationalised rail industry and its union by promoting transport of coal by non-unionised private haulage companies. MacGregor appointed a deputy, Ken Moses (1931–92), to facilitate his plan, which was admitted in March 1984, to close 20 relatively unproductive pits. The planned-for strike commenced and would continue for over a year, resulting in schism of the NUM and defeat for what had been the most powerful trade union in the UK.

In a conversation in 1981 with Joe Gormley one evening in Luxembourg, where we were engaged in obtaining research grants from the European Communities for coal and steel research in Britain, I learnt that he had predicted that this would happen; he believed Scargill's left-wing political views blinded him to some of the realities of the situation that Gormley saw building up. Instead of an inevitable slow, continued decline to death, the industry and its workers were subjected to effective summary execution.[13] By 1994 there were only 15 deep mines left in the UK and the last of these, Kellingley Colliery, closed in 2015. 1985 was the year in which the counterblast to the 1945 Labour Government's introduction of the Welfare State and the post-war political consensus started. Britain has been a less consensual country since then.

One relatively minor consequence of this disruption was MacGregor's appointment of Moses to chair the governing board of the IOM. MacGregor was a Scot with a first class degree in metallurgy and engineering from Glasgow University. He had worked during the war for the government on tank manufacture and thereafter had made his life in the USA, acquiring a reputation for ruthlessness with trade unions. He had been brought back to run the failing British Steel and had made drastic cuts in preparation for privatisation, and he continued in the same vein with the NCB, now called British Coal. Moses had started life as a miner at the coal face and had risen rapidly through managerial posts to become deputy chairman of the board, acquiring a PhD on the way by part-time study. He shared with MacGregor a ruthlessness and dislike of the unions. He was the man whom MacGregor deputed to inform the pits that they were to close. Both men regarded the IOM rather as an 'unproductive pit', although we were able to demonstrate that our work on ergonomics did enable the saving of a great deal of money. But, in fact, we were in the process of quantifying the harmful medical effects of mining, as well as how they could be reduced, and the former cannot have been welcome to them. The prospect of closure of the IOM and possible redundancy of our 120 staff was something that I had prepared for over the previous 12 years, as I recount later. But first, it is necessary to explain what my colleagues' research over my 13 years as director of IOM revealed about coalminers' lung diseases. Did we answer the big questions that had so far eluded others?

Does dust exposure cause chronic obstructive lung disease?

All through the history of lung disease in coal miners there had been a consistent observation – they suffered from bronchitis and breathlessness, and if in life they coughed up black sputum they were likely to die from their lung disease. They also often developed black lesions in their lungs that were seen by pathologists at autopsy and, from the 1920s, by radiologists on chest films. It was easily understandable that doctors should expect that the black lung was the cause of the symptoms; when it became apparent that black lungs and symptoms were not always related to each other save in the case of PMF, there seemed to be only two options: either to assume that the bronchitis and breathlessness were caused by something other than coal dust or that both they and the black lungs were caused by it.

In retrospect it is curious that the medical profession preferred the former explanation and did not seriously consider a third possibility – that dust exposure could add to other causes of lung disease, increasing the risk. Haldane had recorded the presence of bronchitis in coal miners but preferred to think that it was due to hard physical work straining the lungs. Later, when cigarette smoking had become almost universal among adult males in Britain and was shown to cause COPD in a high proportion of smokers, this took the entire blame for causing the disease in miners. There is little doubt that Cochrane was open-minded on the matter, but his studies did not include data on lung function and the data on exposure were limited to preliminary studies in one pit. In his later years he became increasingly sceptical of the IOM results.

Cummins and Gough also suspected that emphysema and bronchitis could be caused by dust exposure, but did not have the epidemiological resources to test the hypothesis. Both were able to show that emphysema was closely associated with dust deposits in the lung, but later Gordon Heppleston (1915–98), working with Gough, decided from his very detailed studies of sections of coal miners' lungs that this was a benign form of emphysema that differed from the usual sort found in smokers.[14] He called it focal emphysema. The IOM research was designed to answer this question, to test the hypothesis that coal exposure is a cause both of pneumoconiosis and also of disability from chronic obstructive lung disease. Estimating exposure on the basis of careful measurements and taking account of smoking habits were central to this.

The first evidence came from Jacobsen's mortality study, though this necessarily relied on death certification, and in coal-mining areas doctors tended to attribute deaths in miners from lung disease to pneumoconiosis, which they did not discriminate from bronchitis. However, this would have reduced the strength of the association of death from bronchitis with dust exposure, so it was reasonably strong evidence. In 1985 the mortality of 25,000 miners followed over 22 years was reported. Those with simple or no pneumoconiosis did not have a greater mortality than that of the general population in the corresponding areas of Britain, whereas those with PMF had between 20% and 350% increase, the risks being higher if the PMF occurred in younger men or if the stage of it was higher (i.e. more extensive radiographically). There was no increase in risk of dying from lung cancer overall, but as expected, smokers had a five-fold greater risk than non-smokers; interestingly, this risk was rather lower in men with pneumoconiosis. Critically, there was a highly significant increased risk of death certified as from chronic bronchitis and emphysema.[11]

Then followed a series of papers on lung function. The first, in 1973, had shown that both smoking and dust exposure were significantly and independently related to reduction in the measure of airflow obstruction, FEV_1. In a subgroup of 1677 men from five pits who had been examined over a 10-year period it was shown that FEV_1 declined at a rate dependent on previous dust exposure irrespective of their smoking habits, the effects of the two appearing to add up.[15] Colin Soutar, who was to succeed me as director of IOM, summarised multiple papers on the subject in 1987, including others from the USA, Australia and Germany using similar methods, all pointing in the same direction.[16] This inverse association of dust exposure and lung function was finally confirmed in a paper in 1988 in which a visiting epidemiologist from the USA, Bill Marine, using IOM data, showed that dust exposure could be associated with clinically significant losses of lung function (Fig. 8.4).[17] The risks approximately doubled if the miner was a smoker, but the relationship with dust exposure was not altered.

How does coal dust cause COPD?

In spite of all this evidence, scepticism remained, notably from Keith Morgan and Archie Cochrane. It was desirable to show how this effect of coal dust could operate. Two mechanisms for narrowed airways seemed likely: inflammation in, and consequent blockage of, small airways (bronchioles) in the lung, or emphysema. As stated above, pathologists had demonstrated the association of emphysema with pneumoconiosis but disagreed on its significance. A hint of its relevance had come from the ingenious study from the MRC's PRU in 1982, which had looked at the lungs of men in South Wales who had died of heart attacks, and showed that those who had worked as miners had more emphysema than those who had not been miners, even after allowing for smoking habit and age.[18] But now we knew that coal dust exposure was associated with the cardinal functional effect of emphysema, reduction in FEV_1, so if the risk of emphysema could be shown to be related to dust exposure, the chain of causation would be complete: dust exposure causes emphysema, which causes loss of lung function, which causes symptoms, disablement and death.

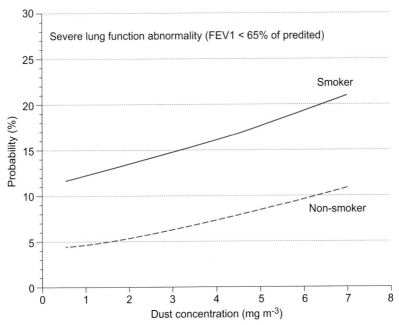

Figure 8.4 Relationship between having an FEV_1 of less than 65% of average normal and daily exposure to coal dust over 40 years underground. Note that the risks from smoking and coal dust are similar and appear to add up. Data from the Institute of Occupational Medicine.

The IOM's field research had allowed us to collect the lungs of a number of deceased miners who had previously taken part in the surveys, and we thus had details of dust exposure, symptoms, smoking habit, X-rays and lung function on most of these men. Anne Ruckley, the IOM physician who led this research, studied the lungs of 342 men from whom full data had been obtained during their working lives. Those who had pneumoconiosis were increasingly likely to have a significant amount of emphysema the greater their coal dust exposure had been and the greater the amount of coal dust found in their lungs (Fig. 8.5).[19] Meanwhile, in the laboratory, Ken Donaldson[20] and his colleague Geraldine Brown were able to show that inhaled coal dust could cause inflammation in the lungs of rats and release of enzymes that were capable of breaking down elastic tissue in the walls of the alveoli.[21] A plausible mechanism whereby coal could cause emphysema had been demonstrated.

Prevention of emphysema and compensation for miners with COPD

The purpose of all this research had been to give guidance to the coal industry on prevention of lung disease in the workers, and the results were used to set standards for dust concentrations in coal mines. Application of these standards, together with X-ray surveillance of all miners and removal to low dust conditions of those with early signs of simple pneumoconiosis, resulted in a substantial and

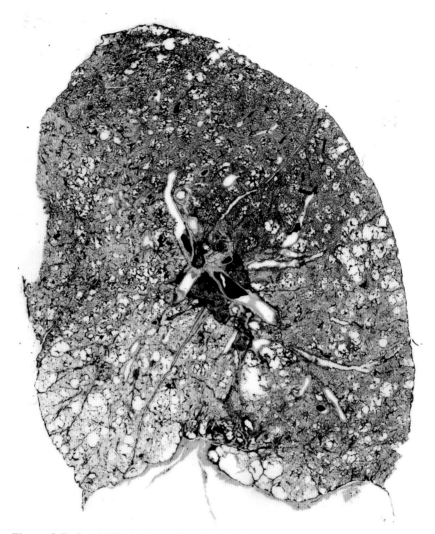

Figure 8.5 Gough-Wentworth section of miner's lung showing white areas of emphysema in lower zones and black deposits of coal in central lymph nodes (photo courtesy of Prof. Ken Donaldson).

progressive fall in the numbers of men with pneumoconiosis in the industry in Britain, and in particular new cases of the serious form, PMF, became very rare (Fig. 8.6). Recommendations were also made to British Coal to reduce the dust standard further in order to reduce risks of bronchitis and emphysema, by then called COPD. However, the industry was contracting rapidly and the prospects for young miners of a career in Britain long enough to cause disease had become remote. Nevertheless, the data were and remain sufficient to be applied with confidence in other countries where coal is mined, and should still play a part in protection of miners.

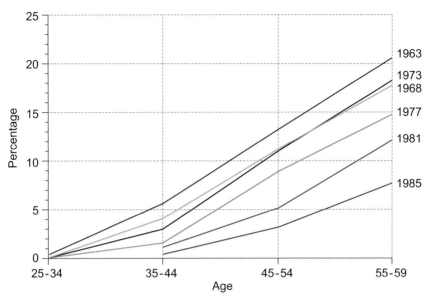

Figure 8.6 Proportion of the UK mining workforce found on regular radiographic surveillance to have pneumoconiosis at different ages, showing progressive decline related to improving dust control. Data from the Institute of Occupational Medicine.

While the research was not intended to inform compensation issues, the data were obviously useful in deciding whether or not COPD should be recognised as an occupational disease in coal miners and in 1993, almost coincident with the announcement of closure of 31 of the 50 remaining pits, the UK Government accepted the recommendation of the Industrial Injuries Advisory Council that chronic bronchitis and emphysema (COPD) should be scheduled as an occupational disease in coal miners. This allowed the payment of compensation to miners on a no-fault basis, irrespective of their smoking habits. The prescription initially required the miner to have been employed underground for 20 years or more and to have early X-ray changes of pneumoconiosis, but this latter requirement was later dropped for practical reasons. The 20-year requirement was to give assurance that the miner had suffered sufficient dust exposure to cause the disease, given the dust levels currently obtaining in the UK industry.

One final twist to this story came in 1999, when a group of disabled miners sued British Coal in the English Civil Courts for negligence in exposing them to dangerous levels of dust. In the course of a very expensive court case, British Coal's lawyers attempted to show that all the research by IOM that the industry had funded, and the results of which they had accepted on the subject was seriously flawed, based on the evidence of Keith Morgan, who had by then become a persistent critic of the methods used. It was, unsurprisingly, a vain effort. The miners won their case and the UK Government then had to agree a method of paying the more generous civil law compensation to many tens of thousands of miners who satisfied criteria for COPD or disabling pneumoconiosis. Less satisfactorily, many doctors and lawyers involved in the process did very well out of the cumbersome method of assessing disablement that the Government deemed necessary.

What is it in coal dust that is toxic?

The one important outstanding question on mining and dust diseases was, what determines the different effects of dust in different mines? In chapter 4 I explained that coal has three components: carbon, the volatile element including various tar chemicals, and ash. The ash includes quartz and a number of silicates, which differ in different coals. Quartz causes a distinctive pathological lesion, concentric rings of fibrous tissue, in the lungs, (Fig. 8.7) recognisable on X-ray by nodules that are rather larger than the usual coal nodules. This disease is called silicosis, as described in chapter 6, which unlike simple CWP tends to be progressive even after dust exposure has ceased. As noted, silicosis can co-exist with CWP in coal miners who drive roads into the face or drill into hard rock, but may also occur if mining conditions force them to cut roof or floor with the coal. In these circumstances, a rapidly progressive pneumoconiosis usually occurs.

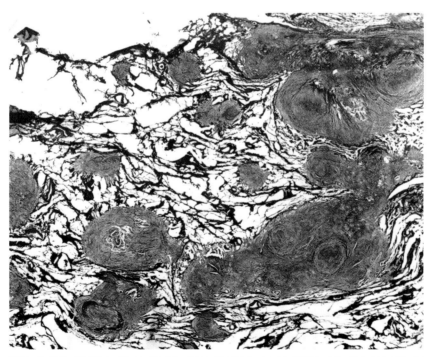

Figure 8.7 Section of lung of coal miner who had worked drilling rock, showing silicotic nodules. On the right they are merging together, to form PMF. The background lung shows emphysema and black coal deposits. (Photo courtesy of Prof. Ken Donaldson.)

One final episode illustrates this and the problems that confronted the British coal industry in its last years, echoing the experience of Fisher in Somerset in 1924 (see p 78). In 1980 I saw a patient in my NHS clinic who worked in a local coal mine south of Edinburgh, one of the oldest in Britain. He was in his 40s and his X-ray film showed PMF. Such a finding was extremely rare in Scottish mines at that time. His pit was one that we were studying in the research, and it was as a result of this that his X-ray abnormalities had been found and that he had been referred to the chest clinic. It should be noted that there is no treatment that will influence the course of pneumoconiosis – the only management that will slow or arrest its progress is prevention of further dust exposure. The pit was on its last legs (it is now a mining museum) and the seam was too narrow for the cutting machine, which had been taking sandstone from the roof as well as coal. He had silicosis, as had several of his colleagues, and this had occurred in spite of the coal dust levels in the air having been well below the mine dust standard.[22] Unfortunately, the standard applied to *coal* dust – the dust in this pit comprised up to 40% *quartz*. Quartz, the cause of silicosis, is many times more dangerous than coal, so that $1mg/m^3$ would be likely to be fatal after a year or two whereas that concentration of coal dust would at that time have been regarded as trivial. The problems in the pit were known to both management and union, but agreement had been reached to keep the pit open to protect the jobs. Various efforts were made to reduce risks, but the effects on the workers proved to be the final nail in the pit's coffin, and it was forced to close. The publication of the details of the episode later made an important contribution towards debate on a preventive standard for quartz in workplaces.

More generally in coal mining the influence of *small* proportions of quartz in coal mine dust is hard to detect in terms of causing radiological or pathological change. Experimental studies have suggested that the toxic effects of quartz particles depend on their surface; freshly fractured particles are more toxic than weathered ones. Confusingly also, the presence of silicate minerals (clay) in the particles of coal dust seems to inhibit the toxicity of any associated quartz. Thus, in general if the percentage of quartz in coal dust is less than 10%, any effect of the dust in causing pneumoconiosis is attributable to the carbon component. Only when the quartz proportion is higher do silicosis-like lesions occur in the lungs and with very high concentrations, as in the case reported above, the consequent disease is properly called silicosis.

The other important factor in the composition of coal dust in causing disease is the rank of the coal, mentioned in chapter 7. This was clearly confirmed by the IOM studies, high-rank coals being more potent in causing CWP and loss of lung function than low-rank ones. The better the coal burns, the more dangerous it is to inhale in dust form. The explanation for this has not been discovered in spite of much speculation; it may reside in some physical properties of the particles such as their number, shape or surface activity that have not been measured systematically. In terms of prevention, rank is an important guide to danger and special efforts towards dust control should be required in higher rank pits.

The final solution; the end of UK deep coal mining and the re-birth of the IOM

The publication of the silicosis article and the arrival of Ian MacGregor as Chairman of the British Coal Board, both in 1981, represented the start of an increasingly troubled relationship between IOM and British Coal. I had seen, within a year of my appointment, that our future was uncertain and that we required to take two strategic actions. The first was to increase efficiency, that is to do more work at lower cost, and the second was to expand the customer base and progressively to move research into other than coal-mining diseases. In terms of efficiency, this meant progressive loss of staff and increases in our rate of publications in scientific journals. This was at first not too difficult, as I had inherited a number of older colleagues who were ready to retire, and natural loss from retirement and movement to other jobs allowed us to reduce from 150 to 100 over the next 10 years. Improvement in publication rate (in my view the best index of productivity for such an institute) was facilitated by the amount of data available to us already.

From my previous post as an NHS consultant, I had been an interested but impartial observer of the Pneumoconiosis Unit in Cardiff and had seen some of the difficulties of an isolated research unit in a city with a medical school. To me, it was essential that IOM should build close relationships with local interests and with other organisations interested in the health of workers. When I arrived at IOM in 1978, 80% of the income came from British Coal with some also from work on asbestos, so most of our efforts were directed towards getting money from other sources, particularly the European Coal and Steel Community and later the EU. But we also gradually built up research in other industries and for government departments and, led by Jim Dodgson and his colleague Alan Bradley, set up a unit to sell consultancies and occupational hygiene services. Over the years our dependence on British Coal lessened to about 25%. We had developed good relationships with the British Health and Safety Executive and our work in diseases unrelated to coal was becoming widely recognised. In particular our research had contributed to development of protective occupational health standards in a number of industries, notably woollen mills, polyvinyl chloride production, shale mining, asbestos use and stone quarrying. We had also developed an interest in more general environmental research issues such as contaminated land and air pollution and, led by Jim Vincent, developed a series of innovative instruments for measurement of airborne particles. In 1989, when we were almost ready to set out on our own, the dénouement came unexpectedly.

Under the pretext of an inspection of the Institute, Ken Moses visited us. I took him round the laboratories to introduce him to the staff, anticipating that this would be followed by a discussion of our options for the future. To my horror, he asked everyone what they planned to do with their redundancy money. One member of staff informed a local newspaper which came up with a banner head-line 'Edinburgh Research Institute to close'. After a harrowing fight, and with the help of Sir Richard Doll [23] who was on our management council, I managed to agree a deal; in return for job cuts among the most senior staff (including myself) and a down-payment by British Coal of what they would have paid us over the next three years, we could set IOM up as a self-funding charitable research insti-tute. To aid this process, we raised a further million pounds from industrial and charitable sources, notably the Colt Foundation,[24] for research into mineral fibres, based on our track record of research into asbestos-related diseases. I named Colin Soutar as my successor and, in 1990, I and the other department heads retired from British Coal's employment. Happily, each of us had ensured that highly capable scientific deputies were ready to take over the running of the IOM and its branches.

By then, British Coal was well on the way to handing its few remaining pits to private companies who would, like their predecessors in the 1940s and before, become responsible for preventing lung disease in their employees. All are now closed and the problem of lung disease has been exported to those countries where coal is still mined. For these countries, the hard lessons in prevention learned in the UK can be applied. Colin Soutar and his colleagues successfully led the newly independent IOM, which continues as a self-funding research charity; in 2019 it will celebrate its 50th anniversary, overtaking the longevity of its parent, the nationalised UK coal industry, which closed in 1997.

Thoughts on the problems of researching industrial disease

Once upon a time it seemed relatively simple to study occupational diseases, and to a limited extent it still is. For example, for a doctor, if someone develops wheezy symptoms at work, and is using or exposed to some chemical known to cause asthma, it is pretty easy to carry out simple tests to discover the cause and then take preven-tive measures. This is the clinical approach. Some things are obvious, but others are more difficult, particularly when it comes to generalising from clinical observations. These chapters have illustrated this; a seemingly simple question, 'does coal mining cause COPD?' has required some 60 or more years of concentrated research by hundreds of researchers on tens of thousands of workers before scientists, regulators and governments are convinced. These complexities arise because most diseases have multiple causes that may add to each other or even interact in altering risks of individuals succumbing. When the coal research started, neither the necessary methods for measuring dust or lung function and for quantifying X-ray changes, nor for statistical assessment and computation of the data, were available.

Adding to the complexity, study of people requires their consent, and while this is relatively straightforward in individual patients, it is less so in large populations of workers. Managers and business owners are naturally wary of scientists coming and finding something nasty in the woodwork of their factory, and trade unions may be suspicious of their motives if managers agree to the study. All studies require both parties in industry as well as the individuals concerned to give their informed consent, and to obtain this requires tactful and open negotiation. An important issue is of course funding; the funder may wish to impose conditions on the researcher with respect to publication. This has always to be resisted; failure to agree means the study should not be done. It is reasonable to explain to the funder that they will receive the results first so they can prepare their response, but it must be agreed that the results are presented to both parties formally within a short period after the research is finished and later published openly. The contract should make this quite clear, to prevent accusation that the funder has in some way influenced the results. In most cases, the suspicion of a hazard has already been aroused, and in that situation management has the option of denying it or taking action to find and reduce it if it exists. The latter course, if supported by the workforce, has obvious moral, industrial relations and often financial advantages.

In summary

The IOM was very fortunate in being supported from 1969 to 1990 by the nationalised British coal industry and having had, up to its last few years, full support from senior management and the coal trade unions. This enabled it to take off where the MRC had left coal research and give convincing answers to the remaining questions: quantification both of the risks of pneumoconiosis from coal dust exposure of different ranks, and also of the somewhat greater risks of developing COPD. As a side effect, these results led to compensation becoming payable to miners with COPD, but the main objective of the research had always been to provide information on which coal-mining industries could set science-based preventive standards. The risks of inhaling coal dust were clearly shown to relate to the rank, or carbon content, of the coal dust, indicating that preventive measures need to be enhanced when high-rank coal is being mined.

This coal research provided a strong base on which the IOM was able to develop its international reputation for impartial and unbiased research, not only into diseases of coal miners but also into other occupational and environmental health issues – a reputation that allowed it to move to becoming a self-funding charity. Part of this diversification involved studies of the risks of cancer in workers, a particularly contentious area, using techniques developed in its studies of miners. One of these took us back to the very origins of the petrochemical industry and the commodity that was to displace King Coal from his throne: oil. This is the subject of the next chapter.

Chapter 9

Oil, the usurper, and industrial cancers

Oil, the fuel that challenged the dominance of coal and has now usurped much of its role in energy production, owes its origins to a geological process analogous to that of coal. It too was formed in association with gas from the decay of plant matter over millions of years, in conditions of low oxygen levels and high temperature and pressure underground. In the case of oil and the associated natural gas, the plant material was mostly marine algae (seaweeds) as opposed to the trees and ferns that were transformed into coal. The discovery of this liquid fossil fuel and its exploitation came much later than that of coal, but its more convenient physical state meant that it quickly took over from coal as a means of enabling transport and in chemical and plastics manufacture. My own interest in the history of oil came about by chance in the early days of exploitation of the North Sea deposits.

One morning in 1980, a thoracic surgical colleague asked me to see one of his patients urgently. I was at the City Hospital in Edinburgh where I had an honorary appointment as a consultant chest physician. The patient had suffered breathing difficulties after removal of part of a lung for what was thought to have been cancer, and was on a respirator. The problem turned out to be straightforward, an attack of asthma, and he had recovered by the time the report on the excised lung came back from the pathologist. Happily, the tumour removed wasn't a cancer; it was pneumoconiosis, a mass of PMF that had mimicked cancer on the X-ray. In explaining this to him I asked which coal mine he had worked in. 'I wasn't a coal miner,' he said, 'I was a shale miner'.

This surprised me, as I knew that the classic textbook on occupational diseases claimed that pneumoconiosis had not occurred in shale mining, by then an obsolete industry in the UK. Since part of his lung had been removed by the surgeon, it was possible to analyse the minerals in it chemically, and this showed that they matched closely those in a sample of shale from the seam the patient had worked in. I made some inquiries in local hospital pathology departments and discovered that several cases of pneumoconiosis had in fact been found in the past but never reported in the medical literature.[1] And I already knew something about the history of the shale oil industry, since it had been a well-described early cause of cancer.

The oil shale industry

If you walk a couple of miles to the west from where I live in Edinburgh, you come across strange grey-coloured spoil heaps, known locally as bings (Fig. 9.1). These are all that remains of the pioneering oil-shale industry of West Lothian that had been founded in 1850 by James Young (1811–83). Young had been born in Glasgow and studied chemistry in night classes at the local Anderson's University, later to become Strathclyde University. Working as an industrial chemist in England he discovered that he could produce oil by distilling a particular type of bituminous coal known as cannel coal, which had been used locally as fuel in fires. He patented this invention and went on a search for suitable oil-bearing mineral deposits, eventually finding them in the banks of the very same River Almond that we walked beside in the first chapter of this book.

Young set up a company to mine the shale deposits, which were extensive to the west of the river, and to crush and distil them to produce paraffin for candles and lights (he acquired the nickname Paraffin Young). Later he produced a fuel that was named petroleum, from the Greek for rock, petra. This was the first commercially produced mineral oil, and Young's company at the time of his death was employing 5000 workers in mining and refining. Young went on to become a Fellow of the Royal Society and a philanthropist, supporting his life-long friend, the explorer and medical missionary David Livingstone, who had been a fellow student at Anderson's University, taking an active part with him in the anti-slavery movement.

The oil shale industry suffered a severe and ultimately fatal blow within a decade of its start when liquid oil was struck in western Pennsylvania in 1859 and an oil rush,

Figure 9.1 All that remains of the shale-mining area west of Edinburgh. Niddrie Castle in the foreground was once a refuge of Mary Queen of Scots after her rescue from Loch Leven. Shale spoil heaps (bings) are seen in the background.

similar to the earlier gold rush, occurred in that state, shortly to spread to Ohio and then to Texas and California. Echoing the discovery and exploitation of coal, oil had been found seeping naturally from the earth in several parts of the world and had been exploited in a minor way for lighting, especially in Azerbaijan. The importance of the Pennsylvania discovery was that this was the first time a large deposit had been found by drilling, using steam power. The oil was transported in barrels, first by boat along the Allegheny River and later by a new railroad to Pittsburgh. The curious use of the term 'barrel' as a unit of oil volume dates from this period. It was a variable quantity, depending on the size of the barrel and the density of the oil, but was eventually standardised (apparently based on the volume of a wine barrel in the reign of Richard II of England in the late fourteenth century!) at 1.59 cubic metres; production in Pennsylvania alone leapt from around 320 cubic metres to 1.6 million cubic metres between 1859 and 1873; in contrast, the annual production of the Scottish oil shale industry in terms of crude oil reached a peak of around 400 thousand cubic metres in 1907.[2]

A further difficulty confronting the Scottish industry was competition with kerosene, which had been patented by Abraham Gesner in Canada in 1854. He also was a chemist who had found that he could produce an oil suitable for lighting by distilling coal, both he and Young having been working on the same idea from the mid-1840s. This competition and the ready availability of both coal and liquid oil caused a sharp fall in the price of oil in the late nineteenth century, the start of a periodic cyclical process that has occurred ever since, popularly called boom and bust. The Scottish industry was kept alive by government support as an indigenous supplier of oil during the two world wars and by its ability to make nitrogenous fertiliser as a by-product, but finally closed in 1962. Among the first beneficiaries of this industry were whales, slaughter of which had previously been the main source of lighting oil. The last was the Institute of Occupational Medicine, for reasons I shall explain.

Mineral oil increased in importance worldwide with the introduction of the internal combustion engine and particularly the mass production of motor cars by Henry Ford in the USA, but the production of shale oil left a legacy; although its mining was relatively safe compared to coal mining, the surface refinery workers making paraffin started to get cancers on their skins. This had been one of a small number of important discoveries in the eighteenth and early nineteenth centuries that led to an understanding of the causes of cancer.

The discovery of the first cause of cancer, soot

Cancer has a deservedly frightening reputation, even nowadays when cures through surgery, radiotherapy and chemotherapy have become increasingly frequent. Indeed, the so-called war on cancer in the public mind and the media seems to have been fought by scientists working in laboratories investigating the biochemical and genetic causes in the hope of finding a cure. However, the greatest advances, less dramatic and hardly publicised, have been in prevention from identifying environmental and lifestyle causes. This story began in 1775, the year of Priestley's discovery of oxygen and the year before the publication of the Wealth of Nations and of James Watt's patent, with a publication by a London surgeon, Percivall Pott (1714–88), entitled *Cancer Scroti*.

Pott was the most notable surgeon of his time. He had been apprenticed at St Bartholomew's Hospital in London and by 1749 had become a senior surgeon on its staff. His name endures in a number of conditions that he described: Pott's fracture of the ankle bones, Pott's spine and Pott's paraplegia (tuberculosis causing damage to the backbone and consequent paralysis) being the best known. However, his description of cancer of the scrotum in chimney sweeping boys had the most lasting and significant consequences, both in prevention and in under-standing of cancer in general. His account,[3] only two pages long, begins:

> Ramazini has written a book *de morbis artificum*; the Colic of Poictou is a well-known distemper; and every body is acquainted with the disorders with which painters, plummers, glaziers, and the workers in white lead, are liable: but there is a disease as peculiar to a certain set of people, which has not, at least to my knowledge, been publickly noticed; I mean chimney-sweepers' cancer.
>
> It is a disease which always makes its first attack on, and its first appearance in, the inferior part of the scrotum; where it produces a superficial, painful, ragged, ill-looking sore, with hard and rising edges: the trade call it the soot-wart. I never saw it under the age of puberty, which is, I suppose, one reason, why it is generally taken, both by patient and surgeon, for venereal, and being treated with mercurial, is thereby soon, and much exasperated...

After describing the dreadful course of the disease and the inefficacy of treatments, including most attempts at surgery, he comments:

> The fate of these people seems singularly hard; in their early infancy, they are most frequently treated with great brutality, and almost starved with cold and hunger; they are thrust up narrow, and sometimes hot chimnies, where they are bruised, burned, and almost suffocated; and when they get to puberty, become peculiarly liable to a most noisome, painful, and fatal disease.

The reader may recall Charles Kingsley's fairy-story of *The Water Babies*, first published in 1863 (Fig. 9.2), which contains subtle messages in support of Darwin's new theory of evolution. The sweeping boy Tom fell down the chimney into a young girl's bedroom, was chased away and found refuge from

Figure 9.2 Tom, the sweeping boy, fleeing from the nanny on his way to becoming a water baby. From Charles Kingsley, *The Water Babies*, illustrated by Warwick Goble, Macmillan and Co. Ltd, London, 1909.

his cruel master, Grimes, in the river, where eventually he joined the water babies. William Blake, Charles Lamb and Charles Dickens also drew attention to the plight of these children, who were even more cruelly treated than those in factories and mines.

In *The Water Babies*, Kingsley contrasts the smoky town with the pleasant countryside; in the town in Pott's time coal was replacing wood and charcoal burning in the grand houses with large chimneys. It is likely to have been coal soot that caused scrotal cancer to appear so commonly in Pott's time in these adolescent sweeping boys. Attempted legislation from the 1770s had had no effect, being widely ignored, and this cancer remained a plague until the early twentieth century, although effective legislation to prohibit the employment of climbing boys had been enacted by Lord Shaftesbury in 1875, exactly a century after Pott's description.

In pointing to a cause of a disease which would previously have been thought to be a consequence of imbalance of the body's humours or some other malign influence, and by drawing attention to Ramazzini's observations on the diseases of workers, Pott started two important lines of thought in the medical profession. First, diseases may be prevented by preventing exposure to the toxic agent; doctors were to learn to look for such agents and to give their patients preventive advice based on factual knowledge. And, secondly, what is it about these agents that causes disease? In terms of prevention, ultimately only legally enforced regulation is effective, the red tape that the owners of businesses so often decry; for this reason legislation, as was the case with chimney sweepers, is slow to come about. In terms of understanding causes, things tend to move more quickly but are dependent on the curiosity of doctors and scientists and the availability of funding. In the case of occupational cancer, matters started to move quickly only from the late nineteenth century with developments in the understanding of organic chemistry as described in chapter 4, and stimulated by the observations of doctors on their patients.

Shale oil cancer

In 1876 an Edinburgh surgeon, Joseph Bell (1837–1911), recorded a number of cases of skin cancer in workers making paraffin in the oil shale industry, thus discovering the second cause of cancer.[4] At the time Bell was renowned for his observational skills, especially with respect to deducing a man's work from examining him physically. The same skills and use of logic led him to play an important part in the development of forensic medicine in Britain. Today Bell is particularly celebrated because one of his assistants, Dr Conan Doyle, later used him as the model for his character Sherlock Holmes.

The risks of skin cancer in the oil shale industry were investigated in the early twentieth century by a general practitioner, Alexander Scott, who was also physician to the oil company founded by Young. He was able to show that cancer occurred usually after about two decades of work involving skin exposure to paraffin and was preceded by the benign skin conditions, dermatitis and warts. He instituted preventive measures, including urging legislation to provide skin protection and washing facilities in the workplace, and introduced a screening programme to detect early skin disease. As a consequence, the risk of skin cancer among the workers was substantially reduced.[5] However, shale oil was used as a lubricant for machinery in the cotton mills of Lancashire, and in 1922 scrotal cancer was described in the operatives known as mule spinners.[6] These men's trousers became contaminated by the oil from persistent close contact with the machinery in an era when washing facilities in workers' homes and in workplaces were primitive at best.

In 1970/80 the Iranian revolution and Iran–Iraq war led to a fall in oil exports from the Middle East and to panic buying of fuel by vehicle owners, with rapidly rising petrol prices. President Carter initiated a movement away from imported oil and the US industry looked towards the extensive oil shale deposits in the Rocky Mountains. By this time the environmental movement had become established and the US government was seeking information on health hazards in the shale oil industry in order to carry out a risk assessment. This coincided with our report of pneumoconiosis in Scottish oil shale miners and speculation that if it caused skin cancer it might also cause lung cancer when inhaled.[1]

The IOM was able to get a grant from the US Department of Energy to study the effects of this industry on the health of those who had worked in it. It proved possible to identify all those who had been employed in the latter years from the records of a 1950 Providence Fund and to study their mortality patterns, the health of surviving workers in the industry, and their oral histories of the social conditions in the industry. This confirmed the risks of pneumoconiosis and showed that skin cancer had indeed been eliminated in the early twentieth century.[7] Happily, no excess risk of lung cancer was found. The research not only provided the US with information on which preventive measures could be based, but also helped to strengthen the reputation of IOM in the USA and to diversify its research programme; it is for that reason that I said that IOM was the final beneficiary of the Scottish oil shale industry.

Lung cancer in the mining industries

Shortly after Bell's description of shale oil cancer, in 1879 two doctors in Germany, F.H. Härting and W. Hesse, first described lung cancer occurring in miners in Schneeberg, the same metal mines that Agricola had studied.[8] The cause was not apparent initially, but after Marie Curie (1867–1934) investigated the radioactive properties of radium and uranium from the same mines, it became apparent that radon gas seeping into the air of the mines was a likely cause. Marie Curie is remembered as the most notable woman in science, winner of Nobel Prizes for both physics and chemistry, and ultimately a victim of an industrial disease since her experiments with radiation led to destruction of her bone marrow and her death from anaemia.

Many subsequent studies confirmed the greatly increased risk of lung cancer among metal miners. It is now established that exposure to radon gas is a general risk in metal mining such as in the tin mines of Cornwall and uranium mines in Canada. In contrast, radon gas levels in British coal mines have been generally low, and as mentioned in the previous chapter, no excess risk of lung cancer has been found in UK coal miners.[9]

The coal tar and coke industries: other causes of cancer

In chapter 4, I described the rapidly increasing importance in the nineteenth century of coke, oil and gas production from coal, both for industrial and, increasingly, domestic uses. The various fractions of oil distilled from coal tar, mainly as a by-product of coke and gas production, had many uses; from lubrication and waterproofing to chemicals, pharmaceuticals and dyes. The discovery of these coal products was overshadowed by the coincident discovery of liquid oil in the USA, since it also could be distilled into multiple fractions with similar applications, and this could be done at much lower cost. Nevertheless, the coal gas industry continued in the USA into the 1940s and in Britain to the 1970s, while the coking industry continues wherever steel is produced.

Industrial skin cancer was first recognised among men making coke in France in the 1870s.[10] The risk was shown to be from contact with tar and was first recognised for compensation among coke workers in Britain in 1907, as was exposure in the shale oil industry. This recognition of coal tar and mineral oils as causes of skin cancer had initiated interest among scientists as to what chemicals in them were responsible and how they worked. From that time, experiments on rabbits and rodents led eventually to discovery of the chemical structures that were responsible.

The leading British researchers in this field were Sir Ernest Kennaway (1881–1958) and his wife Nina. Kennaway was an Oxford scholar who became the leading cancer researcher of his day, and professor of pathology at the Royal Cancer Hospital (now the Royal Marsden) in London. He followed the 1907 experiments of Yamagiwa and Itchikawa in Japan, who had demonstrated the carcinogenicity of coal tar on the skin of rats, by investigating the chemical fractions of the tar and analysing their structure and their potential to cause skin cancers. With his co-workers, he discovered the first two of what became a group of carcinogens called polycyclic aromatic hydrocarbons (PAHs), benzopyrene and dibenzanthracene (Fig. 9.3).[11] These discoveries were fundamental to all further cancer research. The Kennaways were interested broadly in the subject of cancer causation, including lung cancer, which was then becoming more common, and they were instrumental in initiating research with the epidemiologists Sir Austin Bradford Hill and Sir Richard Doll in the 1940s into air pollution and cigarettes as possible causes of

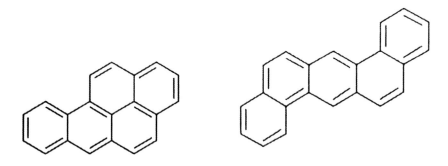

Figure 9.3 Left, Benzo(a)pyrene, and right, dibenz(a,h)anthracene – essentially five benzene rings arranged in various configurations, these being the (**a**) and (**a,h**) one. The letters indicate to chemists the position of the additional rings on the basic three-benzene ring anthracene structure.

cancer.[12] In spite of disabling Parkinson's disease, Ernest Kennaway continued at work long after formal retirement.[13]

Aside from exposure to tar, workers in the coking industry were potentially also exposed to smoke and fumes from the tops of the ovens, where they are charged with coal, and from the side doors through which the coke is extracted. In 1936 there were reports from Japan and England of a suspected excess risk of lung cancer in coke oven workers.[14] The early British report was based on a study of death certification for lung cancer, which as noted above had been rare in 1920 but had risen rapidly in incidence by the 1930s. By 1948 the evidence on coke ovens had grown stronger although, since it affected the relatively small number of people employed in that industry, it could not have explained the overall rise in occurrence of the tumour. This was, of course, shortly to be shown to be strongly related to cigarette smoking. Nevertheless, carcinogens were shown to be present in the oven fumes and these produced tumours in animal experiments.

From the time this became obvious, attempts were made by the industries involved to reduce exposure of their workers while epidemiologists attempted to quantify the risks.[15] Large epidemiological studies continued to show a smaller risk, of the order of a 20% increase among workers on the oven tops at least to the 1980s.[16] Ultimately it became possible, by summarising the results of multiple studies in which measurements and estimates of workers' exposures had been made, to derive a quantitative risk estimate in relation to exposure to the main carcinogen, benzo[a]pyrene.[17] Such estimates provide a basis for effective preventive action.

Benzene and the problems of solvents

In chapter 4 I mentioned the original production of the simple organic chemical, benzene (C_6H_6), from coal tar. The earliest large-scale industrial use of benzene in the nineteenth century was as a solvent for rubber, although it was used on a smaller scale for dissolving inks and paint pigments.[18] Not long after, in 1897, a tragic episode occurred among female workers in a rubber factory in Sweden, nine of whom developed a bleeding disease that was fatal in five. In time it became apparent that this resulted from benzene dissolving in the bone marrow, which is essentially a fatty tissue, and destroying the blood cells that develop there. Thereafter a number of individual cases of leukaemia were tentatively ascribed to benzene poisoning before convincing evidence of a risk in shoe manufacturing industries in Italy became apparent in the period before the Second World War (benzene was a solvent in the glues used). Further episodes were reported from Italy, Turkey and the USA from exposures in the 1950–60 decade. Some readers will recall using benzene over this period in chemical experiments in school as I did; this is now banned and the use of benzene is severely restricted in industry and subject to regulation as a carcinogen.

Not only are benzene and some related solvent chemicals blood carcinogens, but their very property as solvents makes many of them toxic to the brain and nerves, since they also dissolve in the fatty casing of nerves. All solvents need to be handled with care and most are tightly regulated in the West. The commonly used 'trike', trichloroethylene (Fig. 9.4), is now regarded as carcinogenic and several other solvents are currently either banned or under investigation as risk factors for neurological diseases. One of these, carbon disulphide, is mentioned later in this chapter.

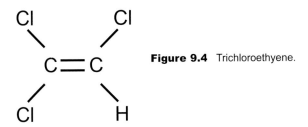

Figure 9.4 Trichloroethyene.

The chemical industries and bladder cancer

In retrospect, it should have been easy, as it was for Pott, to recognise an association between contamination of the skin and subsequent development of cancer. It would have been less obvious to associate the development of tumours of the internal organs with an environmental contaminant. However, in 1895 Ludwig Rehn (1849–1930) reported the cases of four men who had been passing blood in their urine from tumours of the bladder. All were from a local factory employing

500 workers engaged in making dyes based on aniline from coal tar, in the nascent German chemical industry. Such an outbreak of similar diseases in a small population is referred to as a cluster, and is always something that should be taken by doctors and workplace owners as suggestive of a local cause. Rehn was already well known as the first surgeon to have removed a diseased thyroid gland, and was later to become professor of surgery in Frankfurt; his report attracted attention to what became known as aniline cancer.

As with oil and tar cancers, studies on rodents were used to investigate the chemical causes, and eventually it was shown that those responsible were not aniline itself but a group of chemicals derived from it called aromatic amines, notably benzidine and the most potent, 2-naphthylamine. The term amine implies a nitrogen and hydrogen radical attached to the molecule (Fig. 9.5). These amines were present in some processes in rubber production as well as being in widespread use in the organic chemical industry. After the Second World War the British chemical industry started a nationwide epidemiological survey of workers exposed to aromatic amines and showed that they had a frightening risk of death from bladder cancer, on average 30 times higher than did unexposed individuals and almost universal in some heavily exposed groups. These early studies were later discussed by Professor Robert Case of the Chester Beatty Institute (now the Cancer Research Institute) in London, who had been central to them.[19] He pointed to the wide range of countries in which the risk was present and the ways in which it could be reduced. These were by elimination of use of the chemicals and by regular screening of those previously exposed so that early tumours could be treated. He also drew attention to the dangers of the production of toxic chemicals being exported by their companies to countries with weaker controls over hazards, a problem that continues to this day.

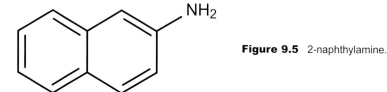

Figure 9.5 2-naphthylamine.

Unexpected consequences: vinyl chloride and carbon disulphide

In chapter 4, I mentioned that both the plastics industry and the organic chemical industry had their origins in the use of coal as their feedstock, although in both cases oil eventually took over as the main supplier. Both industries made revolutionary and beneficial contributions to society. However, to both there has been a downside, not least the contribution to warfare including the production of explosives and poison gases. Few of the great chemical companies can claim innocence in this respect. There has also been a very serious downside to plastics, from their indestructability and therefore their progressive pollution of land and oceans and entry into the food chain. However, the manufacture of several plastics has also been unexpectedly associated with harmful effects on the workers involved, serving as a warning to industry. Notable examples occurred in the manufacture of two of the most ubiquitous man-made materials of all, PVC and artificial cellulose-based fibres.

Vinyl chloride is a simple organic chemical comprising two carbon, three hydrogen and one chlorine atoms, C_2H_3Cl (Fig. 9.6). It was first synthesised by Liebig in the early nineteenth century, and is made by chemical reaction of chlorine, ethylene and air. At room temperature it is a gas that is inflammable and has a marked tendency to explode. Although it was considered as a refrigerant and even as an anaesthetic, these dangerous properties prevented much use until the late 1930s, when it was polymerised to make PVC.

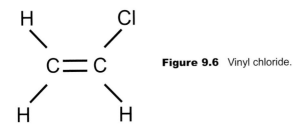

Figure 9.6 Vinyl chloride.

Polymerisation means treating the simple molecule using chemical initiators to break the double bonds between the carbon atoms so they all join up into a plastic material or polymer (Fig. 9.7). From at least the 1940s there were reports by toxicologists of high doses to rodents causing liver damage, and in the 1960s it became apparent that these effects occurred at lower doses and were replicated in PVC production workers, who also started developing painful finger complaints associated with bone damage.[20] By the 1970s liver fibrosis and liver cancer, of a particularly rare type called haemangiosarcoma,[21] had appeared in both experimental rodents and workers exposed to vinyl chloride. Once these were described there was no

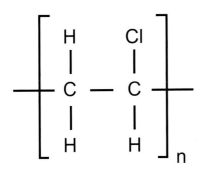

Figure 9.7 Polyvinyl chloride – the n indicates number of repeats of this structure joined together in the polymer.

difficulty attributing them to the chemical, and the manufacturers could no longer disregard the warning signs, leading to much stricter control of the substance in industry. Even such a commonplace and apparently benign material as PVC has left in its trail a story of workers' lives painfully curtailed.[22] The polymer itself is relatively harmless, save when it burns in house fires and as a contributor to the increasing problem of plastic pollution of the oceans and the food chains in them.

The story of artificial silk and cellophane is less well known to the medical profession, but even more tragic in terms of lives ruined, thoroughly documented in a book by Dr Paul Blanc.[23] It relates to a simple chemical, carbon disulphide, comprising one carbon and two sulphur atoms (CS_2 or S=C=S). This was first made in 1796 by heating coal with pyrite, a sulphur-containing ore, by a German chemist, Wilhelm Lampedius. Its structure was elucidated in the early 1800s and it proved to be a very powerful solvent. Its first major application was in electroplating, followed in 1846 by Alexander Parkes's invention of the cold-vulcanisation process for making rubber. In this, the carbon disulphide was used to dissolve sulphur, and then raw latex from the rubber tree was dipped in it to make the firmer and more elastic rubber product with which we are familiar.

Carbon disulphide was assessed as a possible anaesthetic in Edinburgh by the populariser of the properties of chloroform, James Young Simpson. Though effective, it was too unpleasant and malodorous for use in surgery. Within a few years of its introduction into rubber manufacture, reports began to appear of severe acute psychological and neurological reactions in workers exposed to it, notably in Germany and France. Fortunately for rubber workers, alternative methods of vulcanisation became available and carbon disulphide was removed from the process in the early twentieth century. However, in 1892 as the culmination of a search for a method of making artificial silk from cellulose derived from plants, two American scientists, Charles Cross and Edward Bevan, had patented a process of dissolving cellulose in carbon disulphide. This solution could then be forced through apertures to make thread (viscose rayon) or rolled to make cellophane, two transformative industries. Coal remained the feedstock for the carbon disulphide used to make fibres and cellophane through to the 1950s, and the story of these industries and their failure to control the toxic effects on the workers exposed to the vapour of carbon disulphide – acute insanity, suicides and chronic nervous system and eye damage, among others – exemplify dramatically the conflict between the drive for profits and the need to care for the workforce and the environment.

In summary – the hazards of new technologies

The story of coal is one of innovations in technology, and with innovations may come great benefits to humanity. However, the innovator or entrepreneur does not wish for any problems in the use of the technology, and often these problems in the past could not be foreseen. This became apparent towards the end of the nineteenth century with the discovery of cancers in paraffin workers, metal miners and other groups of workers exposed to organic chemicals. Indeed, the development of leukaemia from exposure to benzene occurred before the disease itself was known to medicine. Realisation that work could cause cancer stimulated the development of scientific toxicology, cancer research and all the biomedical research that is now involved in understanding disease processes in general. It is important to point out that, nevertheless, exposure to chemicals at work is relatively uncommon as a direct cause of cancer. Exposure to environmental risks such as cigarette smoke, alcohol, sunlight and a number of infections such as hepatitis B and HIV are much more important causes in terms of numbers of victims. Cancer research has lately focused particularly on predisposition, from analysis of our genomes and especially on the way in which genes interact with environmental factors in causing the diseases to develop – so-called epigenetics. Nevertheless, as we have learnt from this past experience, we have become more cautious about any new technology and are better at attempting to predict and reduce hazard by (sometimes unwelcome) regulation. But now I shall consider two other initially unforeseen hazardous consequences of coal and fossil fuel usage that have had worldwide adverse effects – air pollution and climate change.

Chapter 10

'The inconvenience of the aer and smoake': the story of air pollution

We often hear a warning on the radio that pollution levels in cities are high. To most people in Europe this conjures up a mental picture of cars, buses and lorries belching out exhaust smoke in a congested street. To people of my generation it reminds us of the cold winter days in the 1950s when the fog descended, making our eyes sore and stopping us being able see the houses across the road. To Australians it may mean another forest fire. To someone in California it means a stinging haze that obscures distant views, and to people living near a factory it means the smoke from the chimney. For some it is a stench, for some a sense of irritation of the eyes and nose or a cause of cough, and for others a loss of visibility; but there are many to whom it causes neither physical nor mental discomfort and who might reasonably wonder what all the fuss is about. Yet the announcement on the radio is then accompanied by news items that solemnly inform us that the rise in pollution levels will kill thousands of us, and urging governments to do something about it.

What is this deadly air pollution and just how deadly is it? In our cities it has a history. In chapter 1, I wrote of the early history of coal, of the fleets of ships from the fourteenth century onwards taking sea-coal down the east coast of Britain to London, where it was burned by brewers, dyers, soap and salt makers, lime burners, and hundreds of small trades in their furnaces within the city. Until now, this book has been concerned with the harmful effects of coal and of its useful products on the health of those who worked with them; now we are concerned with the effects of coal combustion on the health of whole populations and eventually of civilisation itself.

In 1661, four years before the Great Plague struck and five years before the City of London was largely consumed by fire, the scholar and diarist John Evelyn (1620–1706), one of the original Fellows of the Royal Society, wrote a tract addressed to King Charles II called *Fumifugium: or the Inconvenience of the Aer and Smoake of London Dissipated*. In spite of his sycophantic dedication – *It was one day as I was walking in Your Majesties Palace at Whitehall (where I have sometimes the honour to refresh myself with the sight of your illustrious presence, which is a joy to your peoples hearts) that a presumptuous smoake issuing from one or two tunnels neer Northumberland-house, and not far from Scotland-yard, did so invade the Court;*

that all the rooms and galleries and places about it were filled and infested with it; and to such a degree, that men could hardly discern one another for the clowd, and none could support, without manifest inconveniency – in spite of this, his blunt exposition of the perils to human health, trees, flowers and buildings from coal smoke went unheeded by the King. The coal continued to come down and be burnt and Evelyn's recommendations for abatement fell on deaf ears.

An early note of the health effects of London fogs was by the physician Sir John Floyer (1649–1734) who in 1698 wrote a detailed account of the disease asthma, from which he suffered himself. He commented: *Any kind of smoak offends the spirits of the asthmatic, and for that reason many of them cannot bear the air of London, whose smoak, like fire itself, irritates their spirits into an expansion.* In a 1933 reprint of *Fumifugium*, the novelist Rose Macaulay drew attention to the then current problems, using Evelyn's words: *Still do the chimneys of London and of the more dreadful cities to the north belch forth noxious and gloomy vapours from their sooty jaws, so that these cities resemble rather the face of Mount Etna, the Court of Vulcan, Stromboli and the suburbs of Hell, than an assembly of rational creatures.* Nevertheless, not everyone complained and some even liked the smogs, speaking of the 'London Peculiar'. Indeed, London became known to artists such as J.M.W. Turner, Claude Monet and Whistler for the impressionistic effects of the clouds of pollution on sunlight.

You can't see air, but you are always aware of its presence. It may be warm or cold, humid or dry. If you run you need to breathe more of it into your lungs, and when you sing you learn to control the way you breathe it in and out. Sometimes you can taste it or smell it. Like water, it may be completely transparent, hazy or opaque, its clarity depending on the amount of particulate matter in it. Sometimes, in spring or summer, it may make your eyes and nose stream with hay fever. Sometimes, in the words of John Evelyn, it may be harmful to the health (Fig. 10.1).

In a sense, everything in air is natural – that is, produced by the planet and its inhabitants, animal, plant and microbial – but we tend to regard some components as pollutants because they imply some harm to us, the dominant species living on the Earth, or to our fellow animals, plants and buildings. In this chapter and the next I propose to take a rather wider view of pollution as simply fluctuations in the respective proportions of the constituents of air that may imply problems for the biological functions of the organisms that depend on it. Everything in the air is in some sort of equilibrium; pollution implies a disturbance of that equilibrium. Even aquatic organisms are part of this as, with a few deep-sea microbial exceptions, they also depend on oxygen or carbon dioxide derived from the atmosphere.

Figure 10.1 The smoky city: Warwick Goble's illustration hinting at something sinister in the air of the city. From Charles Kingsley, *The Water Babies*, illustrated by Warwick Goble, Macmillan and Co. Ltd, London, 1909.

The constituents of air: gases

As discussed in chapter 2, it has been known since the eighteenth century that air is approximately 21% oxygen and 78% nitrogen, but there are small amounts of inert gases such as argon, neon, krypton and xenon present that do not concern us here (though these rarer gases may be exploited – I used radioactive xenon as a marker gas in some of my early research into lung function). There is also a small amount of carbon dioxide present and, as also mentioned before, this is part of the carbon cycle between plants and animals sustaining life on Earth. Its critical importance in climate change is discussed later. The nitrogen gas in the air is relatively inert and you might wonder what it is doing there. From the point of view of animals, it serves the very useful purpose of preventing us being poisoned by oxygen when we breathe air, for if we breathe 100% oxygen for more than a few hours it destroys the delicate tissue of our lungs; oxygen in high concentration is a poison, but in lower concentration is, of course, essential to life.

The proportions of these three important atmospheric gases, oxygen, nitrogen and carbon dioxide, have been relatively stable over long time periods. Other gases are more liable to frequent fluctuations, the main ones being ozone, nitrogen oxides and sulphur dioxide. Of these the most important in human health terms is ozone, which is really oxygen but with three atoms rather than the usual two (O_3). This molecule is less stable than the usual oxygen (O_2) and is therefore very chemically reactive. It is mainly found in the stratosphere, more than 10 miles above the Earth's surface, and has the important property of protecting the Earth from bombardment by cosmic and solar radiation. Were it not there, life could not exist on the planet, which is one of many reasons that make me think it would be inadvisable to enlist on a trip to Mars. Such expeditions can usefully be left to the billionaires who appear to be willing to pay for such a journey. This ozone layer came to public attention in the late 1980s, when scientists working in the Antarctic discovered that a hole had appeared in it, shown to be due to the destruction of ozone by the man-made chemicals known as CFCs, chlorofluorocarbons, used in refrigerants and pressurised aerosol canisters. Fortunately, this discovery led to international preventative action to ban the production of CFCs and the progressive enlargement of the hole ceased, although it is still present above the South Polar Region, perhaps maintained by the actions of other man-made chemicals.[1]

Ozone is also present in low concentrations in the lower atmosphere, the troposphere in which we live, where it is produced and broken down in complex chemical reactions driven by sunlight; it is thus found in higher concentration in sunny climates and in the summer. These reactions are slow, taking hours or even days and, once generated, ozone persists for a similar period, often being carried by wind long distances from its source. Its reactive nature makes it a mild irritant to the eyes, bronchial airways and nose, so may cause symptoms especially in people who are more susceptible as a consequence of their having asthma, allergies or bronchitis. Once generated it is converted by reaction with an oxide of nitrogen, nitric oxide (NO), produced by fuel combustion, into nitrogen dioxide (NO_2) and oxygen (O_3 + $NO \rightarrow NO_2 + O_2$). Thus ozone concentrations tend to be higher in the country than in cities, where vehicle exhausts produce the nitric oxide to convert it to oxygen.

Nitrogen dioxide is another slightly irritant gas produced mainly, as above, from its precursor, nitric oxide, by vehicles and other combustion sources, particularly from diesel and gas burning, but also by volcanic activity. It is not generally known that very high concentrations of NO_2 (and of particles – see below) may be generated by burning gas in your kitchen unless you have an efficient extractor (Fig. 10.2).[2] NO_2 takes part in atmospheric chemical reactions, combining, for example, with ammonia produced by farm animals (derived from their urine) ultimately to form particles of ammonium nitrate. Apart from its effects on the airways of patients with asthma and other chronic lung diseases, its concentration in the air has been shown in numerous studies to relate to risks of heart attacks in older people. This is something of a paradox to which I shall return when I discuss pollution episodes; it does not necessarily mean that the gas causes the deaths. Nitric oxide (NO) itself is not toxic in the concentrations found in the atmosphere; indeed, it is a natural

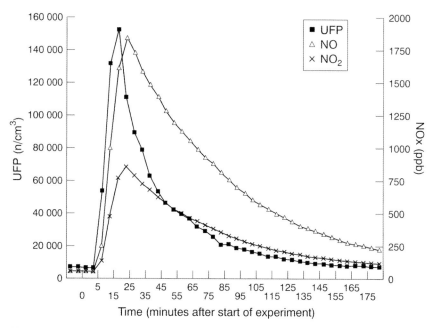

Figure 10.2 Concentrations of ultrafine particles, NO and NO_2 in an unventilated kitchen when four gas rings are burnt for 15 minutes. NO_2 rises to 800ppb then falls over 2–3 hours. The EU standard is 50ppb over 24 hours. Reproduced from: Ultrafine particles and nitrogen oxides generated by gas and electric cooking. Dennekamp, M., Howarth, S., Dick, C.A.J., Cherrie, J.W., Donaldson, K., Seaton, A. *Occupational and Environmental Medicine* 58:511–16, 2001 with permission from BMJ Publishing Group Ltd.

product of some of our cells and plays a role in transmitting messages between cells in our bodies. Whether inhaled nitric oxide might, however, interfere with such natural messaging processes in the lungs is not known but is plausible. It does not persist long in the air, reacting with ozone to form oxygen and NO_2.

Sulphur dioxide (SO_2) is a more irritant gas than nitrogen dioxide and is well established as a cause of sore eyes, cough and wheeze when concentrations rise. Unlike the other gases mentioned so far, it has a distinctive odour described unhelpfully as sulphurous, and in high concentration it is an irritant to the eyes and nose. However, once you have smelt it you will recognise it. If you have visited a volcano while on holiday you will probably know what I mean, as volcanoes are important sources. Coal often contains sulphur in significant quantities, and during the smogs of the 1950s in Britain, in Eastern Europe until recently, and in cities in China and India today, coal burning led to sulphur dioxide being an important part of the pollution mix. Like nitrogen dioxide, sulphur dioxide takes part in chemical reactions in the atmosphere. It dissolves in water droplets to form sulphuric acid particles and combines with ammonia to form ammonium sulphate particles; both of these can also contribute to an irritant pollution episode.

These three gases, ozone, nitrogen dioxide and sulphur dioxide, are the main ones we think of as pollutants, though all are inevitably present in the atmosphere in low concentrations without man's intervention. And they are not alone. The atmosphere is also the recipient of a multitude of other chemical compounds as a result of the natural processes of humans, and indeed, of other animals and plants. For example, when you walk through a pine forest you notice a resinous smell caused by simple organic chemicals, terpenes, released by the trees. Looking out over the countryside on a sunny day you will often see a distant blueish haze produced by chemicals released by vegetation; these take part in chemical reactions in the air to make particles. We do not think of these as pollutants, as they are likely to have natural purposes; some will be aimed at attracting insects, which detect them with their antennae, while others may be natural insecticides. An example of the blurred line between pollutants and natural air chemicals is provided by the case of oilseed rape, the yellow crop used for animal fodder and cooking oil. It also has a distinctive smell from the terpenes it produces, and this is mildly irritating to the lungs of people with asthma.[3] To the bee, it is an attractant leading it to pollen. To the rape, it is a facilitator of its means of reproduction. To people suffering from asthma, it is a nuisance, and they may want the crop to be grown elsewhere. To farmers it must have the smell of profit or they won't grow it.

It is obvious that different industries may pollute the air, as they may pollute the water, with undesirable waste products, and thus many other chemicals may be found in the air derived from local sources. One that has some relevance to coal mining is methane (CH_4). I have already mentioned this toxic and inflammable gas in the context of mining. It is produced when organic matter is acted upon by bacteria in conditions of low oxygen, and may sometimes be seen bubbling out of lakes and ponds. In addition to mining (and fracking), agricultural activities are important sources, as methane is liberated by turning over soil and is present in intestinal gases emitted by animals (including, of course, ourselves). The presence of methane in the ambient atmosphere in low concentrations is now recognised to be of increasing importance, as it is a major contributor to climate change; in this respect it is discussed in the next chapter. Another gas contributed to the atmosphere by farming is ammonia (NH_3), derived from animal urine and the use of fertilisers. This reacts rapidly with nitrate and sulphate radicals in the air to form ammonium nitrate and ammonium sulphate particles, part of the cloud of particles measured by the instruments that weigh or count particles in the air.

Particles

If air is drawn over a sticky tape and examined through a microscope, you see very many tiny particles; the stronger the microscope, the more you will see. Particles are solid or liquid matter of varied shape and many different compositions, and most methods of measuring them in the air do not discriminate between different ones.[4] These all have a tendency to fall through a gas medium, but at different rates depending on their physical characteristics determined by a rule, Stokes' Law. In simple terms, the larger and denser a solid particle, the more quickly it will fall in air (or sink in water). Thus small particles tend to persist in the air, and these are the ones that you can see floating in a beam of sunlight. We can separate particles into biological (pollen grains, fungal spores, bacteria and viruses) and inorganic (minerals and chemicals).

The sticky tape referred to above is actually used to collect pollen and fungal spores in the air in an instrument called a Hurst sampler. Air is drawn into it through an orifice at a known rate, and the particles are caught on a slowly moving tape. They may then be examined under a microscope, identified and counted. For example, the commonly broadcast pollen counts in the hay fever season are derived from these measurements. A similar technique may be used to identify bacteria, which are too small to be identified and discriminated visually; they are allowed to fall onto plastic plates containing a gel of nourishing bacterial or fungal food where each individual germ, when the plate is put in a warm oven, rapidly reproduces to produce a recognisable colony. These sorts of techniques allow scientists to count the numbers of these micro-organisms and to identify them, and this may be used to investigate outbreaks of allergic symptoms or infections in workplaces. You may recall the story of Alexander Fleming's discovery of penicillin; when looking at just such a plate containing bacteria, he saw that their growth was inhibited by fungal growth from the random arrival of a *Penicillium* spore.

Inorganic particles come in a wide range of sizes, from those visible to the naked eye to some as small as viruses, but the ones relevant to human health are those small enough to be breathed in and retained in the lungs. In general, these are less than 10-millionths of a metre (10μm) in diameter, that is, about the same size as the cells in our blood. About the smallest thing you can just see with very keen eyesight and good lighting is a house dust mite, which is about 100μm or 0.1mm long. A particle of less than 10μm has a reasonable chance of getting to the deepest parts of the lung, the alveoli, and staying there where, as I discussed in chapter 5, if it is toxic it may cause damage.

Smaller particles, less than 2.5μm, have an even better chance and that is why particulate air pollution is often reported as PM_{10} or $PM_{2.5}$, particulate matter less than 10 or 2.5μm in diameter. Rather than being counted these particles tend to be weighed, by being drawn through an orifice that selects the size range required and caught on a filter paper, the weight deposited over a given period (usually 24 hours) being expressed in micrograms per cubic metre of air ($μg/m^3$).[5] The most common inorganic particles in the air of towns and cities comprise either soot (carbon) or sulphates and nitrates of ammonia, the former produced by combustion of fossil fuels and the latter by photochemical reactions in the atmosphere. All these particles necessarily carry minute quantities of various other substances on their surfaces, such as metals and polycyclic aromatic hydrocarbons derived from combustion processes.

Many of these inorganic particles are of very small size indeed, less than 0.1μm (100 nanometres) in diameter, and occur in vast numbers in air. They are so small that even in large numbers they make little contribution to the weight of PM_{10}. For this reason and because they may be important in terms of health effects, it is now becoming common to measure $PM_{2.5}$, that is, particulate matter less than 2.5μm in diameter, which better represents them. It is also now possible to get instruments that automatically count these small particles so they may be expressed as the number in every cubic centimetre of air. The relevance of this will become clear when I discuss health effects.

The killer pollution episodes

The river Meuse rises in the Langres plateau in eastern France and meanders gently northwards towards Belgium before entering the Rhine delta to flow into the North Sea, some 925km from its source. At the small but ancient city of Huy it narrows into a valley towards Liège, a stretch of the river sketched by J.M.W. Turner in 1826. The narrower the river, the faster the flow of water, and it would have been the power of the river that attracted industry to the area. In the Middle Ages up to the sixteenth century, cloth, leather and metal working industries became established, flourished and declined. This decline was halted by the Industrial Revolution and in the nineteenth century the Meuse Valley up to Liège became one of the most heavily industrialised areas of Europe. By then, coal had overtaken the river as the dominant source of power.

On 1 December 1930, people living in the villages along the river woke to a dense acrid fog. Some farmers, who remembered a similar fog in 1911, took their cattle up into the hills. Within three days people started choking and complaining of breathlessness, and on 3 December more than 60 people died of chest problems, as did numbers of cattle that had remained at the low level in the valley. It was a national disaster, and suspicion fell upon the local factories, people having recent memories of the use of poison gas in the 1914–18 war. These factories included coke ovens, blast furnaces, metal works, brickworks, power plants, zinc works, a sulphuric acid plant and a fertiliser factory. It is not difficult to imagine what daily conditions were like in the valley.

A judicial inquiry commenced on December 6, led by Jean Firket (1890–1958), professor of pathological anatomy in Lige. Uniquely, the inquiry included a toxi-cologist, meteorologist and industrial chemists.[6] The report of their investigation six months later was able to exclude release of poison gas and suffocation as pos-sible causes, and came to the very plausible conclusion that the deaths were due to a combination of meteorological conditions trapping the usual emissions from the factories and domestic chimneys. These conditions are a winter anticyclone in which cold, still air at ground level is trapped by a layer of warm air above (Fig. 10.3). Domestic and industrial activity requires more heat and thus more emis-sions, leading to smog. It was noted that sulphurous gases were an important part

Figure 10.3 A temperature inversion over the Firth of Forth. The emissions from the distant chimney rise in the cold, still air until they are trapped by the warmer air above and drift to the west.

of those emissions. The committee worked out that the sulphur dioxide released from coal burning would have reacted with water vapour to form sulphuric acid particles and that these, together possibly with soot particles, would have caused the symptoms that the people suffered. Extraordinarily perceptively, the authors suggested that were such an episode to occur in London, the same death rate would result in about 3200 immediate fatalities.[7]

Another river, the Monongahala, got a mention in chapter 5, with its coal barges passing north from the mountains of West Virginia to Andrew Carnegie's steel works in Pittsburgh, Pennsylvania. Twenty miles south of Pittsburgh in the valley formed by the Monongahala lies the small town of Donora, the population of which had risen in the early twentieth century to 13,000 on the strength of industrial developments in the manufacture of steel wire and in zinc refining, the factories powered by coal from the local mines. On 27 October 1948, the town was enveloped in a dense and irritating fog, some 40% of the population suffering respiratory and eye symptoms. By 30 October 17 deaths had occurred and within days another three died, estimated to be six times the expected rate. Again, an investigation suggested that the cause was an anticyclone trapping the usual emissions from the local factories, rather than any sudden increase in emissions or other toxic substances. The emissions commented upon were sulphur dioxide and nitrogen dioxide; at this time particles in ambient air were not generally considered relevant to health. This was soon to change. Donora, which subsequently fell on hard times as part of the so-called United States 'rust belt', is currently reviving somewhat on the back of the shale fracking industry.

Exactly 22 years after the Meuse disaster, Firket's warning was realised. In 1952 Britain had suffered a cold November with a persistent anticyclone and the night of 4–5 December was cold with clear skies. The air was moist and a mist started to form in the morning of the 5th. The inhabitants of London lit their coal fires and the fog became denser as the condensing water in the cold air mixed with smoke and was trapped by warmer air above. On the 6 and 7 December the fog was so dense that it was sometimes impossible to see a yard ahead, and transport ceased. Particle concentrations rose from about $50\mu g/m^3$ to $1.7mg/m^3$ and sulphur dioxide from about 100 to 700 parts per million (Fig. 10.4). The hospitals filled with people, mostly elderly, choking to death, though there was little effect on young people, save possibly for small infants. By the time the weather had changed and the fog had cleared on the 9 December, at least 4000 excess deaths had occurred in the city. The Meteorological Office later estimated that 1000 tonnes of soot and 140 tonnes of hydrochloric acid had been emitted each day, together with sulphur dioxide that was converted into 800 tonnes of sulphuric acid. The acidity of the choking fog was illustrated by 13 deaths among cattle at the annual Smithfield show in the city. Prize cattle, kept in carefully cleaned stalls, died while others in less well attended

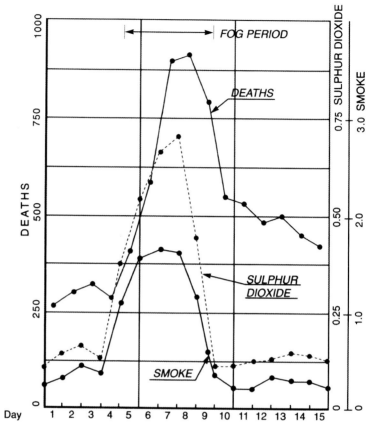

Figure 10.4 Pollution concentrations and daily deaths in the 1952 London smog. Sulphur dioxide is in parts per million and particles (smoke) in mg/m³. Note the rapid rise and slower decline in death rates. Data from the UK Department of Environment, Food, and Rural Affairs.

stalls survived, the ammonia from their urine neutralising the acid of the pollution.[8] In public health terms, this episode was worse even than the cholera epidemic of 1854.

The people living in Britain's cities were used to fog and smoke, but this episode was something altogether different. Some attempts were made to attribute the deaths to influenza by a government opposed to further regulation, but public pressure, largely expressed in the newspapers, led to an inquiry under Sir Hugh Beaver, an engineer and industrialist who was later to found the Guinness Book of Records. The committee's report, which stressed the cost of air pollution in terms of damage to property and loss of productivity as well as effects on health, persuaded a reluctant government to pass the first Clean Air Act of 1956. This was the first nail in the coffin of King Coal: official recognition that coal killed not only miners.

The disappearing problem

Before the 1950s the winter fogs were regarded as a normal part of life, and it occurred to few researchers that they were a threat to life, although it was apparent to hospital doctors and general practitioners that they did affect those unfortunate people who suffered from chronic bronchitis. This disease primarily affected the poor and disadvantaged, and in an era of private practice before the NHS they would have been less likely to attract attention than people with diseases that affected rich and poor alike. One of the few who did pay attention to air pollution was Corbett McDonald (1918–2016). He had qualified in medicine from St Mary's Hospital medical school and had been hooker in their famous rugby team that beat the New Zealand tourists (the All Blacks) before the Second World War. He was later to become professor of occupational epidemiology as a leading researcher into asbestos diseases in both McGill University, Montreal, and the London School of Hygiene. He continued working to an advanced age and still lectured to my post-graduate students when in his 80s. But as a young doctor just out of the Royal Army Medical Corps he had moved to the Harvard School of Public Health on a research fellowship.

In 1951, McDonald and his colleagues wrote a seminal paper explaining the problems of understanding the observations that had repeatedly been made of deaths and illness associated with pollution episodes.[9] In essence, doctors and the public generally shared a mindset that individual diseases have individual causes, whereas in reality all diseases are likely to have multiple causes that interact. Thus it was wholly plausible that several components of air pollution, differing in different places, could interact with individual susceptibilities to cause death or illness. This contrasted with experimental studies that normally looked at high doses of toxic agents in relatively normal people or animals. He called for well-designed epidemiological studies in order to attempt to identify the factors in air pollution responsible for the observed consequences. He then returned to London, by coincidence in time to get practical experience of looking after the victims of the 1952 episode.

Younger people nowadays, if transported back to this era in a time machine, would find life uncomfortable. The streets were still lit by gas, food was rationed and houses were heated by coal fires. The coal was of poor quality, leaving a lot of ash to be cleared in the morning, the best coal being exported to help the balance of trade. Britain was a country impoverished by the war. To keep the house warm and heat the water in the cold winter, fires were banked up with so-called nutty slack, cheap poor-quality coal, which kept them going overnight but produced more smoke than heat. There were relatively few cars, but diesel and even coal-powered trucks were being used to transport goods around the towns, and electric trams were being replaced by diesel buses. In the country, the wheat was threshed by coal-powered machines. The washing, hung out to dry on a line, would gather specs of soot, and a layer of fine black soot would need to be 'damp dusted' from window ledges and furniture every day. We did not realise that we were breathing this same dust into our lungs, nor consider that it might possibly be doing us harm. The Great London Smog of 1952 changed all this; when I attended autopsies at the end of that decade I saw the black pigmentation in the lungs of city dwellers, the melanosis that Läennec had commented on in the early nineteenth century.

The Clean Air Act of 1956 was a direct response of the UK government to the 1952 smog. It introduced smoke control areas in towns and cities and encouraged the use of 'smokeless fuel'. It encouraged movement of power stations and heavy industries out of towns and cities and increased the height of factory chimneys. With its succeeding Acts it was remarkably successful in moving households from burning coal to electric and gas heating, and in Britain central heating became more and more common from the 1970s. Industries slowly switched to cleaner fuels, particularly after the introduction of North Sea gas. The Medical Research Council set up an Air Pollution Unit at St Bartholomew's Hospital medical school in London under Professor Patrick Lawther (1921–2008) to investigate the problems associated with pollution episodes.

Lawther was a medical graduate of St Bartholomew's Medical School and became a consultant physician at the associated hospital, universally known as Bart's. He was to direct the MRC Air Pollution Unit for its entire existence and later worked in toxicology for the Ministry of Defence. The Unit's research into the causes of the associated ill health focused mainly on coal burning and acid smogs, over a period during which, ironically, the smogs themselves largely disappeared from Britain. However, the Unit's interests were wide, including measurement of pollution levels, experimental studies of rodents and people, and lung function studies, the emphasis understandably being mainly on episodes of high pollution and the effects of particles and sulphur dioxide on the lungs. A relationship between particle concentrations and symptoms in people with chronic bronchitis was shown, and this was influential in the World Health Organisation's first formulation of air quality guidance.

The reason for this emphasis is clear from the comment of one original member of the Unit, a chemist, Dr Brian Commins. Speaking about the amount of sulphuric acid in the polluted air during an episode in 1956 and the attempts to alleviate the symptoms of patients with ammonia bottles (to neutralise the acidic pollution), he said of a visit to a ward: *It was the most harrowing experience I've ever had as a young chemist. I'd have been only 25 then. I went to this ward and saw all these patients with terrible breathing problems – you can't imagine what it's like – all fighting for air.*

The mindset was on preventing pollution *episodes*; any insidious effects of day-to-day pollution, much lower than during these episodes, did not seem likely to be significant. Nevertheless, Lawther in 1961 was beginning to think in terms of long-term lower levels of exposure having a role in causing chronic bronchitis. The most influential paper in this respect was published in 1965, in which the authors showed that postal workers in country districts had better lung function than their counterparts in SO_2-polluted London, even when the effects of their smoking was taken into account. This suggested that high exposures to the then acidic air pollution in large cities (primarily due to coal burning) had effects not only on immediate mortality, but also on longer term health of the lungs.[10] But by this time the UK research focus had shifted to cigarette smoking as the primary cause of bronchitis, and the long-term effects of air pollution were not pursued.[11] The campaign to counter the propaganda of the tobacco companies and to persuade people of the dangers of smoking was led by Charles Fletcher, who as director of the Cardiff Pneumoconiosis Unit had been in charge of the research into coal miners' diseases. It was a long campaign, but began to show effects by the late 1960s. A decline in the prevalence of COPD began, and thus a reduction in the numbers of people who were particularly susceptible to the effects of air pollution. Over the same period, domestic coal burning in British towns and cities declined rapidly and was replaced by central heating using the less polluting fuels, gas and oil. The dense winter smogs ceased and statistics on pollution levels in towns and cities, both particulate and gaseous, continued to show a gratifying fall in concentrations (Fig. 10.5).[12] By 1978 it seemed that the air pollution problems in British cities had largely been solved, and the MRC unit closed. However, matters were perceived differently in the USA, and 1979 was

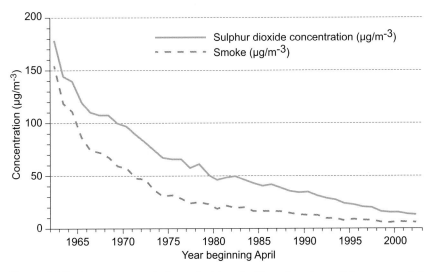

Figure 10.5 Average concentrations of sulphur dioxide and smoke particles measured at over 1200 monitoring stations across the UK. Data from DEFRA brochure, UK Department of Environment, Food, and Rural Affairs.

marked by a paper from distinguished European and Australian scientists suggesting that it was well-nigh impossible to detect effects of air pollution on health at current levels, and a counter claim from the USA that such effects were indeed detectable and important.[13] These differences arose from different attitudes to air pollution. In the USA the emphasis was on setting air quality standards, whereas in Europe it was on reduction of emissions from sources. At that time the methodology of measurement of pollution and the data available to set standards were inadequate to resolve the differences. As time passed, these problems reduced and the Americans were proved to have been right.

The rediscovery of air pollution

There are two intriguing similarities between the story of attitudes towards coal miners' health and safety and that of the effects of air pollution. First, both anthracosis from mining and deaths from air pollution attracted great medical interest for a time, then seemed to disappear, before being rediscovered. Secondly, accidents in mines only attracted serious attention when they occurred *en masse*; the same was true of deaths from air pollution. In both cases it turned out that more deaths occurred as a consequence of small numbers dying on a day-to-day basis. Only deaths in large numbers or in gruesome circumstances attract the attention of the press and politicians; the majority are simply causes of personal sadness. Those many individual personal tragedies are the concern of public health.

The fall in air pollution in Britain after the Clean Air Acts had been dramatic, and it was entirely understandable that for a while British research into the subject ceased after the MRC air pollution unit closed.[14] MRC money comes from the public purse and, as pollution was then regarded as a low health priority, its access to funding disappeared. But the growth of vehicular traffic, particularly in the USA, attracted the attention of environmental groups, and new epidemiological techniques derived from economic analyses became available, as computers became more powerful and better able to deal with large data sets. The first signs that pollution at these relatively low levels was still affecting people's health came from the USA. Alerted by the Donora and London episodes, public health doctors in the USA had noted further pollution episodes associated with excess deaths through the 1960s in New York City. Physicians and epidemiologists started to investigate whether continuous exposure to pollution, rather than acute high-level exposure, might have adverse effects on health. Evidence was already present by 1990 from analyses in individual cities in the USA that mortality increased as particulate pollution rose, irrespective of the acidity of the aerosol or the presence of sulphur dioxide, and it was becoming apparent that particles were the dominant, if not the only, factor in this association.[15]

It was also becoming apparent that if death was a consequence, other less severe effects, such as hospital admission, should be measurable – a point made by David Bates (1922–2006) in 1992.[16] Like Corbett Macdonald and Walter Holland, he had been working as a doctor in hospitals in London during the 1952 pollution episode. He then joined the then current brain drain to North America, in his case to McGill University in Montreal. He was originally interested in lung physiology and published a famous book on the subject with another well-known British expatriate physician, Ronald Christie. He had become especially interested in the effects of pollutants on the lung, and moved to British Columbia, where he studied the effects of ozone on hospital admissions. His concept of coherence between different consequences of single pollutants became very influential in subsequent research worldwide, as has been demonstrated by the huge numbers of papers published since then on everything from stroke to cot death to general practice consultations for infections, all in association with pollution.

Central to this re-awakening of interest in air pollution in the USA were three scientists, Douglas Dockery, Arden Pope and Joel Schwartz. Dockery, originally a physicist and meteorologist, moved to the Harvard School of Public Health where he joined a far-sighted team that in 1976 had established cohorts, groups of carefully selected samples of individuals, in six cities across the USA to be followed prospectively in order to determine effects of air pollution on their health. Critically, data on smoking were obtained, as this was the most important possible confounder of any associations found. In 1993 they published what was to become the most highly cited paper on air pollution, concluding:

> Mortality rates were most strongly associated with cigarette smoking. After adjusting for smoking and other risk factors, we observed statistically significant and robust associations between air pollution and mortality. The adjusted mortality-rate ratio for the most polluted of the cities as compared with the least polluted was 1.26 (95 percent confidence interval, 1.08 to 1.47). Air pollution was positively associated with death from lung cancer and cardiopulmonary disease but not with death from other causes considered together. Mortality was most strongly associated with air pollution with fine particulates, including sulfates.[17]

Removing the technical bits, this suggested strongly that the higher the exposure to particulate pollution, the greater the risk of dying early from heart or lung disease or lung cancer, the risk being about 26% greater in the most polluted city than in the least polluted.

Arden Pope was a co-author of this paper. He also was a convert to epidemiology, being a professor of economics in Brigham Young University in Utah. He took the opportunity of a year-long strike at a local steel works to study the association of changes in particulate pollution with attacks of respiratory illness. The paper concluded:

> This study assessed the association between respiratory hospital admissions and PM_{10} pollution in Utah, Salt Lake, and Cache valleys during April 1985 through March 1989. Utah and Salt Lake valleys had high levels of PM_{10} pollution that violated both the annual and 24-h standards issued by the Environmental Protection Agency (EPA). Much lower PM_{10} levels occurred in the Cache Valley. Utah Valley experienced the intermittent operation of its primary source of PM_{10} pollution: an integrated steel mill. Bronchitis and asthma admissions for preschool-age children were approximately twice as frequent in Utah Valley when the steel mill was operating versus when it was not. Similar differences were not observed in Salt Lake or Cache valleys. Even though Cache Valley had higher smoking rates and lower temperatures in winter than did Utah Valley, per capita bronchitis and asthma admissions for all ages were approximately twice as high in Utah Valley. During the period when the steel mill was closed, differences in per capita admissions between Utah and Cache valleys narrowed considerably. Regression analysis also demonstrated a statistical association between respiratory hospital admissions and PM_{10} pollution. The results suggest that PM_{10} pollution plays a role in the incidence and severity of respiratory disease.[18]

It is difficult to escape another implication of this important paper by an economist – that in addition to adverse effects on health, air pollution must also have impacts on the economies of countries from loss of worker productivity and health service costs. It thus harked back to the report by the Beaver Committee on the 1952 London smog.

The third of the triumvirate, Joel Schwartz, is professor of environmental epidemiology at Harvard School of Public Health. He had first made his name with studies of the effects of lead in petrol, which led to action to remove it, before joining Dockery in his studies of particulate air pollution. He has made very important technical advances in the statistical methods used in analysing the complex data obtained in these studies. His seminal paper with Allan Marcus in 1990 drew attention to the fact that particulate pollution was associated with risks of excess death rates even at concentrations hitherto thought to be trivial.[19] Between them, these three epidemiologists have set the standard for modern studies of air pollution across the world. It is not always realised, thinking back to Louis Pasteur, that many of the greatest advances in medicine have come from non-medically qualified scientists. The critical thinking and mathematical skills of these three epidemiologists revolutionised our understanding of environmental effects on health.

It is now apparent, from hundreds of studies in different countries, that increases in the levels of particles in the air are associated with increased deaths and hospitalisations from heart attack, chronic obstructive pulmonary disease, and stroke, mostly though not exclusively among the elderly and smokers. These associations occur even at quite low levels of pollution, suggesting strongly that in individuals they are not the only cause, but act upon vulnerable people to deliver the *coup de grâce*. There is convincing evidence that the irritant gas sulphur dioxide, which is now largely controlled in the West, increases this effect. What is not so clear is whether oxides of nitrogen, at the generally rather low levels occurring today in UK or US cities, have similar effects, or whether they are simply markers of the traffic-related particles with which they correlate closely.

The particle paradox

My interest in air pollution arose from previous work on occupational lung diseases, especially among miners, and came about in a strange way. I was, in 1992, working at Aberdeen University when I got a call from a senior medical officer at the Department of Health, Dr Robert Maynard, claiming to have been taught by me in Cardiff years before. He had started his medical career as a physiologist before moving to the study of poisons and then switched to work in the UK Health Department. He had noted that the UK had fallen well behind the USA in air pollution research and, with minimal support, he had worked to revive interest in the subject, especially in quantification of the effects of different pollutants. He asked me if I would consider chairing a new government advisory committee on air pollution. At the time I knew little about the subject, but he was persuasive and the opportunity of doing something useful was attractive. The committee, entitled the Expert Panel on Air Quality Standards (EPAQS), fortunately comprised real experts from different disciplines who very quickly taught me all I needed to know about the subject. Recommendations to Government for some standards were relatively easy to determine, for example the gases, where there was plenty of human and animal evidence on which to base them, and which in general did not have significant effects at very low exposures. But then we came to particles.

Traditionally in Britain, particles had been measured by drawing air through filter paper, this being converted to units of micrograms per cubic metre ($\mu g/m^3$) by colourimetry, measuring the density of its blackness. The method was widely known as British Black Smoke. During the 1952 smog these levels had risen to several $1000\mu g/m^3$, that is, several milligrams (mg). Such levels were familiar to those of us studying coal miners, who would often be exposed to $10mg/m^3$ (that is, $10,000\mu g/m^3$) or more of coal dust in the 1950s. However, it had become apparent that deaths were occurring in London at levels of soot particles in the air that were over 100 times lower than those in coal mines where the coal mine dust, also primarily carbon, was known to cause pneumoconiosis but not acute episodes of death. Moreover, the critically important analysis by Schwartz and Marcus of daily deaths in London over the winters from 1958 to 1972, published in 1990, had shown that the deaths had increased in a linear fashion at extraordinarily low concentrations, below 100µg/m^3.[19] In the same analysis, they had effectively ruled out the possibilities that this association was likely to be due to changes in temperature, humidity or sulphur dioxide. Thus, tiny doses of particles were associated with excess numbers of deaths and, from a toxicological point of view, those particles were primarily carbon, which would have been regarded as inert. It did not seem plausible.

The explanation offered at the time was that in the general population there were many people with serious lung disease who were on the brink of death and who were tipped over the cliff edge by a relatively small amount of inflammation in their lungs as a consequence of something in these combustion particles. While this may well have been an adequate explanation of the excess deaths among people with chronic lung disease, it was difficult to understand how the effects could occur at such low doses, and it ignored one important fact, brought out by many studies and summarised in an article by Dockery and Pope in 1994;[20] though the association of pollution with deaths was strongest for deaths from lung disease, the greatest numbers of deaths occurring in pollution episodes were from heart attacks. In their discussion, the authors wrote: *The results of epidemiologic studies of acute effects of particulate air pollution, particularly those describing associations with cardiovascular mortality, have been called into question because of the lack of a biologically plausible mechanism.*

After discussing various possible explanations, all relating to a lung mechanism, including misdiagnosis, they concluded that although the mechanism was unclear, the strength of the evidence was such that it was very unlikely that the associations were artefactual. The next year, at a meeting in the USA, I had breakfast with Arden Pope.

Small particles, big effects: pollution and heart attacks

The question of plausibility was a serious one, since all epidemiological associations are but uncertain evidence of causation; there always remains doubt as to whether some other factor that has not been examined might be the cause of the outcomes observed. In 1990, temperature changes, misdiagnosis of cause of death, even unremarked infections such as influenza, were being suggested as causes of the apparent excess deaths from heart attack. But some interesting facts had not entered into the discussion among scientists interested in air pollution.

The first fact was something that had been known for many decades, indeed back to the nineteenth century: *air pollution particles are predominantly very small indeed*, in the range known as nanometres in diameter (Fig. 10.6). A nanometre

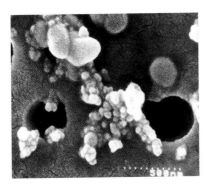

Figure 10.6 Electron microscope photograph of air pollution particles on a filter paper. The scale at the bottom left represents 500 nanometres or 0.5 micrometres. The small particles are in the size range of bacteria. (Photo courtesy of Prof. Roy Richards.)

is one thousand millionth of a metre, or a millionth of a millimetre, the size range of bacteria, viruses and large molecules. Particles of this size had recently become of interest to toxicologists, notably Gunther Oberdörster in Rochester University, New York State. He had qualified as a doctor of veterinary medicine in Germany in 1964, but since the 1970s had been in the USA studying experimentally the effects of inhaled particles on the lung, stemming from an interest in pneumoconioses. The second fact was that in the early 1990s he and his colleagues had published a series of papers that showed *particles in the nanometre size range were handled differently by lungs than larger-sized particles, being retained more efficiently and generally causing more inflammation.*[21]

The third fact was that a protein in the blood, *fibrinogen, which is a necessary factor in the formation of a blood clot, fluctuates in concentration and is generally higher in the winter months.* It was known to increase in response to infections, but this did not explain the seasonal variations. It was also known to cardiologists that increased levels of fibrinogen in the blood were associated with increased risks of heart attack.

A fourth fact, however, was well known to air pollution scientists: *pollution particles cause inflammation in the lungs.* This was accepted as an adequate explanation of the deaths of people with chronic lung disease during air pollution episodes. Finally, a fifth fact that needed to be taken into account was that *nanometre-sized particles were expected on physical principles to diffuse readily indoors.* The importance of this is that most people, especially the elderly who are most at risk of heart disease, spend most of their time indoors, where they might expect to be protected from traffic pollution. In fact we are not, and we later showed that, in general, concentrations of nanoparticles are about half indoors than in the street outside (Fig. 10.7).[22]

Putting all these facts together led to an obvious conclusion and a plausible answer to the question, why do such tiny doses of combustion particles cause apparently well older people to drop dead of heart attacks? Because the lung's role in defence against infection causes it to mount an inflammatory reaction when

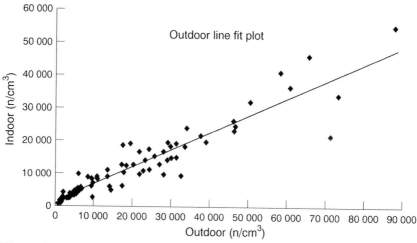

Figure 10.7 Relation between air pollution particle numbers, outdoor and indoor measured simultaneously, showing their close correlation. Reproduced from: Acute respiratory effects of particles: mass or number? Osansunya, T., Prescott, G., Seaton, A. *Occupational and Environmental Medicine* 58:154–59. With permission from BMA Publishing Group Ltd.

bacteria are inhaled; the more bacteria the greater the threat and the stronger the response. In simple terms, during pollution episodes, the lung is fooled by all these nanoparticles into thinking there is a massive attack by bacteria, and the resulting inflammation causes the blood to become more coagulable, increasing the risk of a clot blocking the narrow coronary arteries that supply oxygenated blood to the heart muscle – a coronary thrombosis.

The paper written to propose this hypothesis in 1995 removed the main reason for denying that pollution could cause heart attacks and sparked off a lot of further research.[23] It was then that I first met Arden Pope at breakfast. I wrongly assumed him to be medically qualified and was surprised when he quizzed me on inflammation and blood clotting, until he explained that he was an economist! Subsequent research, notably by Hans-Erich Wichmann and Annette Peters in Germany and the USA, has shown that air pollution is indeed associated with multiple cardiac events and changes in inflammatory markers in the blood, although there are interesting questions still left to answer about the mechanisms involved.[24]

Some thoughts on death: sadness and statistics

There is a small postscript to this story. In December 1999 I left the clean air of Edinburgh to chair a meeting in central London. As luck would have it, there was a temperature inversion and on the second day there I had a heart attack that was shown to be caused by a clot blocking one of my coronary arteries. Thanks to the skill of my cardiological colleagues, who unblocked the artery and inserted a stent, I lived to tell this tale. I could not help noticing that the catheters they used to insert the stent were made of plastic; the stent itself was made of steel mesh. My life had perhaps been close to being ended by fossil fuel combustion particles and had been saved by materials derived from the use of the same fossil fuels.

As a doctor I was necessarily familiar with death, an event we all encounter in our relatives and friends as well as our patients, and which we know we shall meet ourselves. As clinical doctors we learn to help people cope with the process of dying, but try to control our own emotional responses in dealing with the inevitable sadness of the event in our patients. We thus distance ourselves from this most important aspect of death until it affects someone we are close to. As scientists we take this to an extreme, and look on deaths as statistics, as numbers or rates. In doing so, we try to understand the multiple factors that led to this outcome, why someone died at that particular moment in his or her life, what environmental and genetic factors led to that consequence. As clinical doctors we do something rather different: when we ask what caused the death, we mean 'what was the disease that delivered the final blow?' Be it a heart attack (which we call a myocardial infarct), a stroke (cerebral haemorrhage or cerebral infarct), respiratory failure, kidney failure or whatever, this disease is then recorded as a cause of death for statisticians to pore over.

This explains why statisticians were able to attribute death to air pollution, but why this attribution met with a certain incredulity among clinical doctors. No doctor would ever have put 'air pollution' on a death certificate as a cause of death; they simply record the name of the organ that failed, the immediate medical cause. Had I died with the century in December 1999, my death would rightly have been attributed to myocardial infarction. However, it would also have contributed one more death to the hundreds of others that occurred on that day in that city and, if the hundreds were significantly more than expected, mine might have been one of the excess deaths that day attributable to air pollution. But nobody could have pointed me out as an individual who died from air pollution rather than any of the others that day. It could only be said that some of us who died would not have done so that day had it not been for the high levels of pollution at the time. Each one of us presented a more open goal to the airborne particles than did those who did not suffer a heart attack; we would have been on average older, fatter, less active, more likely to have smoked, and more genetically predisposed to the disease. And for the unlucky ones in the lottery of life, the pollution gave the final shove. In case you are wondering, I was not overweight, had never smoked, was physically active and careful with my diet; I did, however, have a strong family history of heart disease and had been exposed to considerable air pollution in most of my younger life.

What then is the point of all this research on air pollution? It is not intended to make a diagnosis of the cause of death in individuals, but rather it is aimed at determining some of the various factors that contribute in a population to shortening of active life, either from premature death or disablement, thus enabling and justifying action to prevent and delay such occurrences. Harking back to John Snow and William Budd in chapter 3, once you know the risk factors, you can take action to reduce them across the population. It accords with the Utilitarian principle of Jeremy Bentham, of doing the greatest good to the greatest number. Although how this should be done is much argued over by politicians, it also accords with the principle first pronounced by the Roman politician, Cicero, who wrote that the public health should be the first responsibility of government. Modern governments always at least pay lip service to this precept.[25]

Where has the research led?

The early twenty-first century has seen an enormous outpouring of international research into the effects of air pollution. In this account I am not concerned with ozone, which bears little relationship with coal but is an important air pollutant, especially in sunny climates. The research has concentrated primarily on particles and nitrogen dioxide, both of which, in many studies, show relationships to heart and lung deaths. Nitrogen dioxide correlates quite closely with counts of nanoparticles (Fig. 10.8). Many studies have shown associations of NO_2 with heart attacks but few

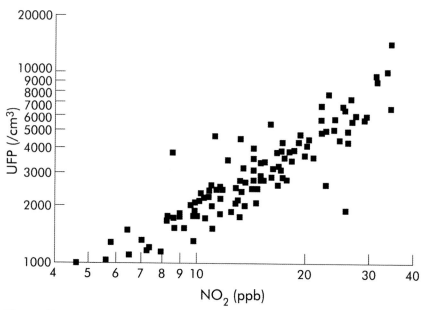

Figure 10.8 Relation between mean outdoor counts of nanoparticles and NO_2 (in parts per billion), representing 6 months of continuous side-by-side measurements and showing a close correlation. Reproduced from: Hypothesis: ill health associated with low concentrations of nitrogen dioxide – an effect of ultrafine particles. Seaton, A., Dennekamp, M., *Thorax* 58:1012–15, 2003 with permission from BMJ Publishing Group Ltd.

of these have measured particle numbers concurrently, so whether the association is actually due to nanoparticles rather than NO_2 remains unresolved. To me this certainly seems plausible.[26]

The best established outcomes of the research so far have been reviewed and may be summarised: [27, 28]

- A $10\mu g/m^3$ rise in PM_{10} is associated with an approximately 1% increase in deaths from heart attack and around 3% increase in deaths from lung disease. Similar rises in non-fatal heart and lung episodes also occur. These episodes occur predominantly in people past retirement age, especially those with COPD.

- Such rises in pollution are associated with smaller increases in deaths from non-haemorrhagic stroke (usually a clot forming or lodging in a brain artery).

- Long-term exposure to air pollution increases the risks of death from heart or lung disease by about 6–7% for a difference in average exposure of $10\mu g/m^3$. The implication of this is considerable, since it implies that unavoidable exposure to air pollution from combustion sources over our lives is a risk factor for the damage to arteries that we call atheroma,[29] a condition that is not confined to the arteries supplying the heart, but that could affect any organ. There is experimental support for air pollution having a role in atheroma formation.

- In several studies air pollution has also been related to an increased risk of lung cancer.[30] This association is consistent with the fact that polycyclic aromatic hydrocarbons are present in low concentration in urban air pollution and may be found on the surfaces of combustion particles.

- Most of the toxicity of the particles seems to reside in the smaller size range including the nanoparticles.

- Some evidence suggests that risks of death and hospitalisation from episodes of respiratory infection in young children are related to air pollution episodes.

- Some evidence suggests that air pollution is a weak risk factor for cot death and for pulmonary embolism.

- There is some evidence from highly polluted cities that pollution exposure may be a risk factor for dementia and even for impaired brain development in the young. These are hypotheses worth testing, not established facts.[31]

- It is very unlikely that ambient air pollution is a significant factor in *causing* asthma in the general population, although those with asthma are more likely to suffer attacks on high air pollution days.

Most studies contributing to these statements point to particulate pollution, though often it is impossible to disentangle contributions from irritant gases, especially nitrogen dioxide, and it is likely that these co-pollutants would, if anything, add to the effects of particles. Sulphur dioxide and acidic aerosols derived from it were important irritants in the London smogs and undoubtedly contributed to lung deaths, but are less important pollutants in the West now that high-sulphur coal is little used. The overwhelming message is that it is very likely that particulate air pollution brings the date of death from heart and lung disease forward in a population in proportion to the concentrations in the air, and that reducing these will reduce this effect. It is likely that two mechanisms contribute to the cardiac effects – life-long exposure to pollution (often, but not necessarily, combined with cigarette smoking) which sets the scene by promoting atheroma, and short-term rises in pollution, which cause the inflammation that precipitates rupture of the atheromatous plaque. In the case of lung deaths it is probable that years of exposure to cigarette smoke are usually necessary to render the lungs susceptible to the final blow from a short-term rise in pollution. All these common causes of death – heart, lung and stroke – are of course also, and even more strongly, related to cigarette smoking, which adds plausibility to the findings, since cigarette smoke is also a mixture of toxic particles and gases.

The exile of King Coal and air pollution today

By the end of the twentieth century the writing was on the wall for coal as a fuel in Britain, as its use declined in its main applications of gas and electricity generation and steel production. Deep mining was costly, and importation from cheaper sources, including Australia and South America, and use of coal from British surface mines made more economic sense, but even these markets were threatened by oil. Continuing studies of air pollution, now primarily caused by vehicle combustion of petrol and diesel, showed that the effects first clearly demonstrated after the 1952 London episode still persisted in spite of the lower concentrations. However, because of those reduced levels the absolute numbers affected would have fallen.

The influence of a reduction in death rates from cessation of domestic coal burning, for example, was shown in a landmark paper from Dublin where coal burning was banned in 1991. After the ban, particulate air pollution declined by an average 35.5μg/m³, respiratory death rates fell by 15% and cardiac deaths fell by 10%. The authors calculated that there were about 359 fewer cardio-respiratory deaths per year or a fall from 5.75 to 4.94 deaths for every 1000 person-years.[32] The implication of this is that far fewer people in the UK and western cities generally are affected by air pollution today than were two or three decades ago, and any acute effects on children are relatively small in public health terms. In terms of mortality, air pollution in general across the UK is now a much less important cause of death than other social factors relating to relative poverty,[33] although its effects add to these in further disadvantaging the poor.

The morbid effects of air pollution are now clearly established to be largely a consequence of combustion of fossil fuel and the production of small particles and perhaps associated gases. The presence of polycyclic aromatic hydrocarbons in the same mixture, chemicals that are often carried on the surface of small combustion-generated particles, may explain the possible risks of lung cancer from air pollution. However, the elimination of coal as a factor has not altered the risk; the same amount of particulate pollution seems to be associated with the same consequences, be it from coal fires or vehicle engines. Thus in the UK and USA the exile of King Coal has led to responsibility now being attributable largely to his successor, King Oil.

Petrol or diesel? Neither

The influence of vehicles on pollution was, of course, obvious in the 1990s when I was involved with advising on air quality standards. People love their cars, and after the defeat of the miners' strike the UK road haulage industry thrived at the expense of the railways. Legislation was introduced in the UK and Europe, first to stop lead (a brain poison) being used in petrol, then to compel new petrol cars to be fitted with catalytic converters to remove gases other than carbon dioxide and water vapour from their emissions. European regulations progressively forced vehicle manufacturers to reduce their carbon dioxide emissions, while some car manufacturers illegally introduced devices in their engines to prevent inspections detecting their failure to comply. In the 1990s I had often been asked by journalists about the effects of air pollution, and a frequent question was 'What should I drive, a petrol or a diesel car?' Diesel was and remains a more efficient fuel in terms of miles per gallon, but produces more particles and gases. On the other hand, petrol produces less pollution but more carbon dioxide per mile driven. So my answer was 'Both are bad for the environment and people. But what is important is to consider your needs and drive as little as possible and as economically as possible.' Since then I have watched the sizes of cars increasing, the number of short journeys increasing, and a chaotic deregulation of public transport in the UK. And in cities drivers still accelerate furiously when the lights go green, only to slam on the brakes at the next stop. Fortunately the penny has now dropped, and fossil fuel-powered vehicles will soon become a thing of the past, at least in the West. Pollution and climate change are caused by people, you and me, and we need to be aware of this.

The exile of King Coal

King Coal ruled Britain and the declining British Empire for over two centuries until King Oil deposed him from the 1960s, but he has fought a dogged battle ever since to retain power in other territories. Economic growth depends on the availability of resources, energy and intelligence, and so long as countries aspire to improve the conditions of their people they will strive for growth. This is a theme for the following chapters, but I shall introduce it here with reference to pollution. In simple terms, pollution is an index of waste. We speak about a machine as being 40% efficient – 40% of the energy produced is devoted to the mechanical objective required, 60% being wasted in producing (for example) heat, noise and smoke. In general, the entrepreneur will wish his production processes to be efficient and his products to be competitive, though not necessarily efficient in the longer term – if they wear out more quickly he may sell more for a while until he is found out. But air pollution can be taken as a measure of the overall efficiency of industry in a city or country. In general, low-income countries have low levels of pollution, which rise as they industrialise and then decline as they realise the inefficiency of their processes and take action to reduce emissions. This is what happened in the industrial West but was delayed in the Communist era in Eastern Europe, and in China, India and South America. Much of Africa is still in the pre-industrial era. The problem of air pollution in the major cities of those nations most needing growth has been compounded by the availability of both cheap coal and oil and the ubiquity of motor transport on which growth also depends.

Only now, as the 2020s approach, has the world started to wake up to the realisation that the future of our civilisation depends on resolving the conflict between climate change and perpetual economic growth, as we shall see in the following chapters.

Chapter 11

The story of a changing climate: the scientific discoveries

It was not until the 1980s that I became aware of growing concerns about a rise in the Earth's temperature, mainly as a consequence of a very public dispute as to whether it was happening. On one side there seemed to be scientists exploring ice in the polar regions and investigating deep-ocean currents, while on the other were politicians of a conservative or reactionary inclination and big business. In the 1990s I had the good fortune to meet a number of scientists interested in air pollution and, as recounted above, became occupied for a decade in helping to advise Government on air quality standards. During this decade I started to wonder about the science behind the concept of climate change and behind the most prominent advocate of its dangers, ex-Vice President of the USA, Al Gore, who had been a student of a leading climate scientist, Roger Revelle (see later). To me, the simplest approach to understanding scientific discoveries is to attempt to trace the history of their development.

The discovery of gases and the carbon cycle

The story can be said to have started with the discovery of the gases during the eighteenth century, first with Joseph Black and Daniel Rutherford and the discovery of carbon dioxide and nitrogen. Black's two most important advances in science were in showing the value of weighing chemicals before and after reactions and the discovery of latent heat. Both were fundamental to the development of chemistry and physics and pointed the way to our future understanding of the climate and of man's influence on it.

Black was born in Bordeaux in 1728 of Irish and Scottish parents and was schooled in Belfast, then attended both Glasgow and Edinburgh Universities. He was influenced by William Cullen (1710–90) who was professor of medicine in both universities successively, and who had led the world by introducing chemistry to the medical curriculum in Scotland. While still a student, Black had designed accurate scales (or a balance) that could weigh very small amounts of chemical matter, one that is still familiar to all students of science. In 1754, when he first published his experiments, chemistry was believed to be based on five 'principles' which could react with each other; water, salt, metals, earth and fire. In investigating the properties of common medicines used for treating indigestion and kidney stones, magnesia and lime water, Black showed that when they were treated with acids they gave off a gas and lost weight. At the time, only one gas was known, air. He also showed that the material left after the gas was released was capable of gaining weight and reconstituting itself by absorption of air. He called the gas that came off 'fixed air', that is, air that has been fixed to the substance, and thus showed that there was another principle involved in chemical reactions. Later he showed that his fixed air did not support life and thus differed from air itself. In the first decade of the nineteenth century, the English scientist John Dalton (1766–1844) proposed his atomic theory, and thenceforward it became possible to describe Black's experiments in molecular terms, such as:

Magnesium carbonate plus sulphuric acid produces magnesium sulphate, water, and carbon dioxide – $MgCO_3 + H_2SO_4 \rightarrow MgSO_4 + H_2O + CO_2$. You will note that the number of atoms of magnesium (1), carbon (1), oxygen (7), hydrogen (2), and sulphur (1), and therefore the weights of the reagents, balance on either side of the equation.

Following Black's discovery of carbon dioxide, Joseph Priestley discovered the gas that supported combustion and Antoine Lavoisier named it oxygen and showed its role in animal metabolism. Another important gas with respect to climate change, methane, mentioned previously as a hazard in mines called fire damp, was discovered bubbling out of Lake Maggiore in 1776 by the Italian, Alessandro Volta, who was to become known for his contributions to understanding electricity. Priestley also, in his experiments on gases, showed that plants were able to refresh the life-supporting factor in air (oxygen) after it had been removed by animal respiration, thus leading the way to the later discovery of photosynthesis. In the 1850s Justus von Liebig in Germany began to understand the chemical mechanisms of photosynthesis and the importance of nitrogen to plant metabolism. The two processes, animal respiration and plant photosynthesis, are fundamental to understanding life on Earth and the process of climate change. They are extremely complex in biochemical terms but can be summarised very simply:[1]

- Animals and many micro-organisms take in oxygen and use it to break down glucose in their cells to produce the energy they need for growth and reproduction, producing as waste products carbon dioxide and water. This process is, of course, analogous to combustion in which fire breaks down carbon-containing material to produce the same CO_2 and water. Both fire and animal respiration are central to climate change.
- Plants and some micro-organisms take in carbon dioxide and water and use the energy from sunlight to convert them into glucose, from which they derive their energy for growth and reproduction. Thus vegetation is central to preventing climate change.

The reactions may be summarised chemically as something that sounds like a cookery recipe, and in fact cookery is a specialised form of chemistry in which we use heat to transform chemicals such as fats, proteins, salt and sugars to make them more edible and tastier. The important reaction is:

Six parts of carbon dioxide plus six parts of water equal one part of sugar and six of oxygen.

$$6CO_2 + 6H_2O \leftrightarrow C_6H_{12}O_6 + 6O_2$$

Sunlight powers this reaction towards the right in plants, whereas in animals oxygen is the fuel to drive the chemicals that release the energy from breaking down sugar. The circulation of carbon in this way between animals and plants is known as the Carbon Cycle.

When you think about it, the origins of life on this planet are very likely to have been derived from combinations of chemicals under the influence of sunlight or geothermal energy in the ocean depths (although it is just possible that some micro-organisms may have developed elsewhere in space and been deposited here). The original chemicals are sure to have included carbon dioxide, nitrogen and water, and the energy mostly to have been derived from sunlight. Ever since then, there has been a balance between animal and plant life, between oxygen and carbon dioxide in the atmosphere and the oceans, disturbed by periods of turbulence of the Earth's surface and of climate change. These periods have, in the past, been associated with extinctions of many life forms; for example, the end-Permian event 250 million years ago when 90% of all species including the ubiquitous trilobite disappeared, and the end-Cretaceous, marked by the disappearance of dinosaurs 66 million years ago. Five such eras of loss of many species have been identified from the fossil record in the past history of the Earth. Could current climate change be leading to a sixth? This is a prospect biologists are now contemplating.

Carbon dioxide: the unseen pollutant

As schoolchildren we learned about carbon dioxide and that the purpose of breathing was to eliminate it while supplying us with oxygen. In my early medical practice I saw many of those patients that had so affected Dr Brian Commins in his account of a visit to a hospital ward in chapter 10, blue and gasping for breath. They were usually in the end stages of COPD, the smoker's lung disease, in what we call ventilatory failure – their lungs were so damaged that they could no longer get rid of the CO_2 accumulating in their bloodstream. Their veins were dilated, their lips blue, and they were tremulous and sometimes confused. Sometimes we were able to get them through the episode by use of stimulant medicines, and later with the help of mechanical ventilators (what the media refer to as life-support machines), to come back next winter as the viral infections or temperature inversions came round again; often we failed. As medical students we had all taken part in experiments wherein we breathed air in and out of a large bag to demonstrate the effects of carbon dioxide. Gradually the oxygen in air is used up and replaced by carbon dioxide, the body responds by increasing the depth and rate of breathing and you begin to feel dizzy and stop the experiment. In patients with COPD, the carbon dioxide in their blood accumulates and turns it acid, their brain blood vessels dilate and start to squash the brain, and untreated, they fall mercifully unconscious and die.

When, in the 1970s, I read reports of rises in the levels of carbon dioxide in the atmosphere, my first thoughts were that they could never rise to the sort of levels that would cause us to die in this way of carbon dioxide poisoning. The concentrations were so low: in comparison to oxygen (21%) carbon dioxide comprised only 0.03% of the air. Not only did this seem an insignificant amount, but it was many times lower than that required to stop carbon dioxide diffusing out of the blood in the lungs and away in our breath.[2] But I was naïve. This was not the problem, and eventually I managed to find out what the worry was about. The story again went back to the end of the eighteenth century.

Carbon dioxide and temperature

So far in this book we have looked back on the eighteenth century as a period of enlightenment, of the flourishing of science and technology, and of close relationships between their practitioners across Europe and even over the Atlantic Ocean. This is a British viewpoint; things were very different in France, where the mid-eighteenth century was a period of political and social chaos, revolution and mass killings from civil war and judicial execution up until the rise of Napoleon.[3] Three characters in this story were caught up in this. René Laënnec, the inventor of the stethoscope who puzzled over black lungs, witnessed the Terror in Nantes as a teenager. Antoine Lavoisier, who named and investigated oxygen, fell foul of Robespierre from his activities in tax gathering, and lost his head on the guillotine, after his judge is famously, but probably apocryphally, said to have stated that *La République n'a pas besoin de savants ni de chimistes* (The Republic needs neither intellectuals nor chemists), a sentiment that is not unfamiliar to students of politics in twenty-first century Britain and the USA. A more reasoned view came from Lavoisier's friend, the mathematician Joseph Lagrange, who commented that: *Il ne leur a fallu qu'un moment pour faire tomber cette tête, et cent années peut-être ne suffiront pas pour en reproduire une semblable,* (It took them only a moment to cut off this head, and perhaps a hundred years would not suffice to reproduce one like it).

The third character is less well known, but his name lives on in the world of statisticians; the mathematician Jean-Baptiste Joseph Fourier (1768–1830). He, like Lavoisier, was imprisoned during the Terror in 1794 but had the good luck to be freed after Robespierre lost his own head. He went on in 1827 to make the observation, using mercury thermometers, that the atmosphere absorbs less of the radiant heat coming towards Earth from the Sun than of the reflected radiation passing in the other direction. He concluded that this was responsible for the temperature being much higher on the planet than it would have been were there no atmosphere.[4]

The importance of this observation is that it explains why our planet, of all in our Solar System, is inhabited by living organisms; the atmosphere traps some of the heat from the Sun, giving us a gift of its energy. Understandably this attracted much scientific attention. What was it in the atmosphere that had this effect? The Irishman, John Tyndall (1820–93), had come to England as a surveyor for the Ordnance Survey, but his interest in science took him into the study of physics in Germany, where he also became a noted mountaineer. Observations on glaciers and awareness of geologists' speculations about the existence of past ice ages (see below) led directly to his later studies of atmospheric gases. He returned as a schoolteacher and then as professor at the London Royal Institution. He became one of Britain's greatest scientists and teachers of science. Among his many achievements was to show by experiment in 1859 that this heat-trapping effect of the atmosphere was a consequence of the absorption of radiant heat (infra-red radiation) by compound gases, while elemental gases such as oxygen and nitrogen had no such effect. He showed carbon dioxide to be 750 times more effective at trapping radiation than air or oxygen at the same pressure. Other compound gases such as water vapour, nitrous oxide (N_2O) and sulphuric acid vapour (H_2SO_4) were even more potent.

In discussing his findings, Tyndall concluded that the quality of radiation was important – the compound gases were transparent to light (a high-frequency form of radiation) but absorbed the lower-frequency heat radiation to different degrees, whereas the same chemicals in elemental form were transparent to both.[5] He conjectured that relatively small changes in the proportions of compound gases such as carbon dioxide in the atmosphere could have been responsible for past changes in climate, and he stated in a lecture in 1860:

> The bearing of this experiment upon the action of planetary atmospheres is obvious. The solar heat possesses … the power of crossing an atmosphere; but, and when the heat is absorbed by the planet, it is so changed in quality that the rays emanating from the planet cannot get with the same freedom back into space. Thus the atmosphere admits of the entrance of the solar heat but checks its exit; the result is a tendency to accumulate heat at the surface of the planet.[6]

This was the key observation – the heat reflected back into space was different from that coming in, and was thus absorbed by these compound gases. What was happening was that the incoming thermal radiation was of higher frequency (ultra-violet) than that reflected out (infra-red) and thus more readily trapped.

It was already known that glass had a similar property, a matter of common observation, though not entirely explicable on the basis of interference with reflected radiation. He also described the eponymous Tyndall effect, the scattering of light by nanometre-sized particles, which forms the basis of apparatus for measuring such particles in air pollution, as mentioned in the previous chapter. In 1884 an American scientist investigating the climate of mountains, S.P. Langley, found ancient dead tree trunks lying way above the vegetation level on Mount Whitney. He was studying solar radiation and speculated, following Tyndall: *I cannot doubt that the changes in those conditions of the atmosphere's transmissibility of heat, which we have climbed into these altitudes to study, are connected with the answer to this riddle.* The riddle was the cause of climatic change in the history of the planet, the explanation of Ice Ages.[7] Indeed, from Tyndall's time and into the late twentieth century, climate research was driven in large part by the geological knowledge that ice ages had occurred repeatedly in the past and would be likely to recur in the future. What climatic changes had been responsible? Could we take any action to avoid another one? Tyndall's researches suggested that carbon dioxide might afford an answer to these questions.

The sinks for carbon dioxide: vegetation and the oceans

This greenhouse theory was challenged (and still sometimes is) on the basis that the concentrations of CO_2 were too low to have a significant effect, and that all the effect was attributable to water vapour in the atmosphere, the vapour of H_2O also being a compound gas. However, in Sweden the scientist Svente Arrhenius (1859–1927) who was to win a Nobel prize for his work on electrolysis, published a paper in 1896 entitled *On the influence of the carbonic acid in the air upon the temperature on the ground*, in which he used the relatively sparse data then available to calculate that a 50% rise in CO_2 would raise the Earth's temperature by 3.5°C.[8] He plausibly suggested that this would act to prevent a future ice age and that the rise in temperature and CO_2 would be beneficial to agriculture in the more northern latitudes. In other words, the CO_2 would act as a fertiliser and the planet's response would be an increase in the capacity of vegetation to absorb the excess being produced.

Thereafter, interest in the issue declined until just before the Second World War, when a combustion engineer and amateur meteorologist, Guy Callendar (1898–1954) made detailed calculations that showed an association between measurements of CO_2 in the atmosphere and the rising temperature over the previous 100 years.[9] He explained this on the basis of his calculations of emissions of CO_2 and its absorptive effect on radiation, and recognised that the oceans, as well as vegetation, were responsible for absorbing CO_2 from the air. His original hypothesis was that their capacity to absorb the excess CO_2 being produced by the industrialised world would be reduced when the sea water became more acidic and saturated with the gas. To return to simple chemistry, water plus CO_2 forms the relatively weak carbonic acid, $H_2O + CO_2 \rightarrow H_2CO_3$. He argued that although the volume of the oceans is vast, only the surface layers down to 100 feet or so are available to take up the gas which then acidifies it; this upper layer is then only very slowly equilibrated with the deeper ocean, and thus is capable of becoming saturated with carbon dioxide. This would mean that it could no longer be relied upon to take up more from the atmosphere.

Callendar's estimate of the effect of CO_2 on temperature, using the much more reliable data then available in 1938, was about half that which Arrhenius had calculated. He was optimistic (remember that he was a combustion engineer) that improvements in engine efficiency would limit the rises in CO_2 accompanying industrialisation, and this led him to calculate that CO_2 levels would rise perhaps from 290ppm to 330ppm in the twenty-first century and to 360ppm in the twenty-second century. He ended his article by saying:

> In conclusion, it may be said that the combustion of fossil fuel, whether it be peat from the surface or oil from 10,000 feet below, is likely to prove beneficial to mankind in several ways, apart from the provision of heat and power. For example, the above mentioned small increases in temperature would be important at the northern margin of cultivation, and the growth of favourably situated plants is directly proportional to the carbon dioxide pressure. In any case, the return of the deadly glaciers should be delayed indefinitely. As regards the reserves of fuel these would be sufficient to give at least ten times the amount of carbon dioxide there is in the air at present.

This was an extraordinary paper and well ahead of its time. However, it illustrates the dangers of prediction which, as ever, is dependent on the assumptions made and the quality of the data, and is subject to the intervention of chance or unforeseen events. His assumption of a limited increase in CO_2 production was a serious underestimate, even though his calculation of the likely consequential rise in atmospheric CO_2 was fortunately closer to what turned out to be reality than had been that of Arrhenius. But overall, the message from these early studies was reassuring. Rises in CO_2 were likely but would probably be beneficial, even though the absorptive capacity of the two sinks, vegetation and the oceans, might be exceeded.

When the Second World War intervened, other things occupied scientific minds for almost two decades, but out of military research were to come key advances that gave a significant boost to climate research, and different conclusions were to be drawn. They involved study of the Earth's ice caps, the oceans, and nuclear physics.

Climate science moves to the United States

By the mid-twentieth century it had been established that CO_2 was a greenhouse gas and had the potential to raise global temperatures, though it was far from clear whether this was indeed happening or, if it were, whether it would be beneficial or dangerous to mankind. Most people at that time regarded the whole matter as of little importance, as we worried about the build-up of nuclear weapons and the confrontation between the USA and the USSR, the Cold War. Recovering from the devastation of the Second World War, Europe embarked on a long, slow process of rebuilding, culminating eventually in the European Common Market and the European Union. The world's population resumed its exponential growth with its accompanying need for energy and food. The United States, financially unharmed and industrially boosted by the war in spite of having made (with the USSR) a decisive contribution to defeat of the German and Japanese forces, established itself as the centre of scientific research. Whereas before the war many ambitious scientists would have spent some formative years working in other European countries, notably Germany, from the 1950s they increasingly went to the USA and many stayed there. Part of the reason for this flowering of American scientific research was the Cold War. For example, the development of nuclear weapons required research also on measurement of fallout of nuclear material, while the development of submarine technology required better understanding of oceanic currents; much military funding was therefore directed towards scientific research, and this was to have important secondary consequences in understanding climate change.

Discovering the past climate[10]

James Hutton, mentioned in chapter 1, first explained the significance of rock formations and deduced from them the great age of the Earth, leading to the science of geology. While his theories attracted enmity from the religious establishment of the late eighteenth century, they were eventually accepted. Hutton in 1797 was among the first to speculate that erratic stones in Alpine valleys had been conveyed to their resting place by glaciers in past ages (see Fig. 1.1).[11] Further studies of these rocks and the geology of mountain valleys led eventually to the understanding that there had been past ice ages, a concept that became apparent with further geological and palaeontological discoveries over the nineteenth century. The competing theory from the eighteenth century, based on the biblical account of Creation, was that the apparently otherwise inexplicable distribution of fossils in rocks and fossil bones discovered in caves was a consequence of redistribution by the great Flood.

The relatively simple view that the strata, having been laid down, had remained in place save when disturbed by earthquakes and volcanic eruptions, became accepted dogma until it was challenged in the 1920s by the German meteorologist, Alfred Wegener (1880–1930), with his theory of continental drift. These two geological concepts, continental drift and periods of glaciation, stimulated much research into glaciers and Wegener led several expeditions to Greenland, site of the northern hemisphere's greatest glacier, in the course of which he established the first Arctic weather station. After being wounded while serving in the First World War and having several near escapes in the Arctic, he finally perished during his 1930 expedition. His theory was largely dismissed by the scientific establishment until the 1960s, when evidence from the magnetic characteristics of rocks showed that he had been correct and the movement of the Earth's tectonic plates was confirmed.

Central to the development of climate change science were two extraordinary advances in technology. First was the development of techniques for drilling through thousands of feet of ice in the polar regions and obtaining cores representing compressed snow deposited every year for hundreds of millennia. The second was development of instruments capable of measuring minute quantities of chemical isotopes by mass spectrometry. The story of the relevance of ice to understanding of climate change can be said to have started in a little-known episode in the Second World War. Greenland had been a colony of Denmark since the end of the First World War. It had always been sparsely populated by Inuit people, who were joined by Viking settlers during the medieval warm period from the seventh century until the return of the ice in the late 1300s. During the Second World War Denmark was overcome by the Nazis who, recognising the importance of the Greenland weather station to their Atlantic warships, invaded and took it over. For the same strategic reason, when the USA entered the war they in turn took it over, and eventually it became an important part of NATO defences in the Cold War.

One member of Wegener's 1930 expedition who survived, Ernst Sorge, had made measurements of the density of ice at different depths in a 15-metre-deep pit he had dug in the Greenland glacier, and had shown that these changes in density allowed identification of annual deposits of snow, rather as tree rings identify different years of growth (Fig. 11.1). In the 1950s this was taken further by a Swiss-born US citizen, the geologist Henri Bader, by drilling into the ice to extract

Figure 11.1 Annual ice deposits shown clearly in a Peruvian glacier. http://beyondpenguins.ehe.osu.edu/issue/science-at-the-poles/unlocking-the-climate-history-captured-in-ice. (Photo by Lonnie G. Thompson.)

cores. He had been appointed director of the US Snow, Ice and Permafrost Laboratory of the Army Corps of Engineers. His insight was to recognise that these cores in polar regions, where there was no summer melting of snow fall, would hold traces of ancient atmospheric conditions. He speculated that investigation of their content of such airborne matter as pollen grains, industrial pollutants and volcanic ash could give information on past climate and factors that might have influenced it. While his aim was scientific, it required justification in terms that would satisfy the military, and this was provided by the fact that such cores would also reveal radioactive traces of past nuclear activity in an era of testing of atomic bombs. The research also received an impetus from international agreement to coordinate polar research in the 1957 International Geophysical Year, which facilitated collaboration between European and American scientists interested in research in glaciers.

The ice and its trapped air

In the early 1950s climate change was not a concern of scientists, and in spite of Callendar's paper the general view was that the oceans had plenty of capacity to absorb all the surplus CO_2 being produced. A change in this attitude towards the effects of increasing industrial emissions on the natural environment started with the publication of Rachel Carson's book *Silent Spring* in 1962, which led to the concept of environmentalism and Green politics. A major scientific advance came from the introduction of carbon dating, a technique for which Willard Libby of the University of Chicago won his Nobel Prize in 1960. In essence, the element carbon comprises two stable isotopes, ^{12}C and ^{13}C, and a mildly radioactive one, ^{14}C. [12] The latter gradually decays by loss of a proton, half disappearing over 5730 years. All living creatures take in carbon from the air, directly as plants or indirectly by eating plants or other animals in their food chain, and this has the ratio of stable and radioactive carbon present in the atmosphere at the time they were alive. After death the ratio alters as the ^{14}C decays, so appropriately adjusted measurements of the radiation emitted from samples (material of known date from Egyptian mummies and tree rings of known age were used for validation) would give their age. Fossil fuels contain none, as all the ^{14}C has decayed after about 50,000 years. The ratio in the atmosphere is stable, since ^{14}C is constantly made in the upper atmosphere by bombardment of nitrogen by protons created by cosmic radiation.

This technique, coupled with the invention of highly sensitive mass spectrometers, opened the way to accurate dating of past ages and of the concentrations of gases in the atmosphere at different periods of time. The mass spectrometer passes substances through a magnetic field and separates atoms according to their atomic weight. Willi Dansgaard (1922–2011) of the University of Copenhagen showed that the oxygen atoms in the water of snow contained a ratio of the isotopes, heavy ^{17}O in relation to the common ^{16}O, that varied with the seasonal temperature at the time the water froze into snow. Using this technique to identify the seasonal change in polar temperatures, he was able to determine the age of ice cores when other methods of counting density change of ice deposits became impossible beyond about 50,000 years. Chester Langway, an American geologist who had joined Bader's team, and a Swiss geophysicist, Hans Oeschger, became interested in using this on the deeper ice cores being drilled in the 1970s, particularly in the deeper Antarctic ice. Crucially, they also noted that the bubbles in the ice would contain air trapped as the water froze and therefore contained tiny samples of the atmosphere at that time (Fig. 11.2).

As analytical methods improved, these bubbles were to become of great significance. Drilling methods had by then enabled Greenland cores to be taken down to bedrock through over 4500 feet of ice. Techniques to measure the gases in those bubbles of trapped air were developed, essentially by careful cleaning of the cores from different depths, crushing them in a vacuum, and analysing the gas emitted. Remarkably, in collaboration with Dansgaard, these scientists eventually measured the atmosphere and temperature back over 100,000 years, beyond the start of the last ice age.

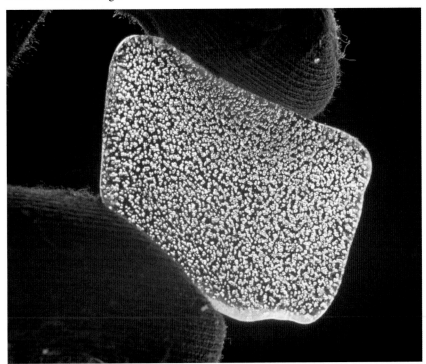

Figure 11.2 Air bubbles trapped in Antarctic ice. Source: CSIRO, CC BY 3.0, https://commons.wikimedia.org/w/index.php?curid=35439336.

From the late 1950s cores had also been drilled in the Antarctic, notably at the Russian base, allowing the research to go even deeper in time, and in 2008 the researchers were able to publish, in the journal *Nature*, data on the Earth's atmosphere and temperature as far back as 800,000 years ago.[13] These showed multiple saw-tooth fluctuations in CO_2 concentrations, the low points coinciding with glacial periods. The most striking feature of these data is a periodicity of approximately 100,000 years in the concentrations of gas in the atmosphere coinciding with that of the temperature, raising the question of cause and effect. Does the fall in temperature cause the fall in carbon dioxide or vice versa, or does some other factor, such as the Earth's axis in relation to the Sun, cause both?[14] But one thing is very clear: the peaks of CO_2 before each decline to a glacial period were no higher than 300ppm, the levels in Arrhenius's time at the end of the nineteenth century. As we shall see, the present levels of carbon dioxide, over 400ppm, are at least 100ppm higher than they have been in the past 800,000 years. A second feature is that within the record there is evidence of shorter-term periods of rapid and probably biologically significant cooling and warming, such as that during the medieval warm period and the more recent Little Ice Age. The causes of these remain subject to speculation.

The oceanic sink

Knowledge of the presence of large-scale oceanic currents dates back to the early days of exploration. The first description of the Gulf Stream was in 1513 by the Spanish explorer Juan Ponce de León, who is said to have discovered Florida. It was first mapped by Benjamin Franklin, whose curiosity was aroused by the different times it took trading ships to cross the Atlantic in opposite directions. It starts in the Gulf of Mexico and, initially driven by winds, arches across the Atlantic towards northern Europe, bringing relatively warm water towards the British Isles and Norway. For the main part of this course it appears to be driven by its temperature and salinity, which determine its density. As heat evaporates the water, it becomes cooler and more saline, and thus denser as it travels north, where it meets cold water from melting ice. This causes it to sink to the deep and flow back southwards towards the Antarctic and join the great currents that circulate round the Pacific and Indian oceans. The heat in its surface waters has an important influence on the climate of Western Europe, rather as the fluctuations in water temperatures in the southern Pacific Ocean known as El Niño and La Niña influence climate on both sides of that ocean. However, the great oceanic circulation including the Gulf Stream appears to have been stable, at least over the period of recorded history. Whether this might change or has changed in the past is a matter of speculation.

Perhaps the key figure in understanding climate change, Roger Revelle (1909–1991), was a Californian graduate in geology and oceanography, with a particular interest in oceanic chemistry. He had served in the US Navy in the Second World War, in relation to submarine detection, reaching the rank of Commander. In 1951 he was appointed director of the then rather small Scripps Institution of Oceanography. Having a talent for organisation and for attracting collaborations from experts across different disciplines, he rapidly built this up into a world-leading research institute. With his oceanographic experience in the US Navy, he was in a good place to attract research funding on oceanic chemistry and currents. In 1957, he recruited two colleagues, Hans Suess and Harmon Craig, who were interested in carbon dating, respectively of tree rings and of oceanic waters.

Revelle was aware of Callendar's work and was interested in understanding how carbon dioxide moved between the air and the oceans, using measurements of ^{14}C. With his colleagues he was able to show that molecules of CO_2 in the air took about a month to be absorbed into the sea, and then took about a century to pass into the deep ocean. He also showed that the simple process of acidification outlined above was much more complicated, as the ocean has chemical methods of neutralising acid with calcium salts (technically called buffering) that meant much of the gas absorbed was passed back out into the air. The oceans seemed not to be as effective at removing the gas from the atmosphere as had been supposed. By the late 1950s, scientists were beginning to take account of the rise in the Earth's population and the rapidly increasing industrial activity that accompanied it. Following the publication of *Silent Spring*, the public had also become aware of loss of rainforest and other vegetation, the other main destination (or sink) for atmospheric CO_2. It became essential to measure the changes in atmospheric CO_2 accurately. One scientist who had been doing this, studying annual fluctuations in relation to plant photosynthesis at the California Institute of Technology, was Charles David Keeling (1928–2005).

Revelle recruited Keeling to the Scripps laboratory, and from 1958 they started the background atmospheric measurements of CO_2 from a base on an extinct volcano, Mauna Loa, in Hawaii. Within a few years they had established the presence of annual seasonal fluctuations on a background of a steadily rising concentration. The seasonal fluctuations, with a rise in winter, relate to the growing period of vegetation in the northern hemisphere. These measurements have continued ever since in spite of periodic threats to their financial support, and constitute what has become known as the Keeling curve (Fig. 11.3).[15, 16] Using isotopic analysis of the CO_2 they have also been able to show that about 45% of the CO_2 in the air is derived from fossil fuel combustion – almost as much as from animal respiration.

It was now possible to combine the recent measurements with those of ancient atmospheres into a composite graphic (Fig. 11.4). Brought up to date, the current concentrations are at a peak one-third higher than at any time in the past 800,000 years, back past some eight troughs representing ice ages, and well beyond the

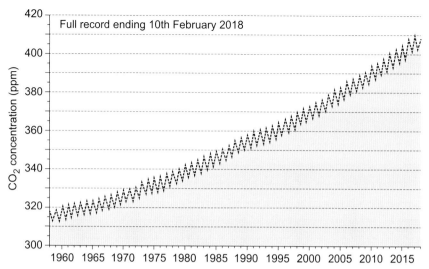

Figure 11.3 The Keeling curve; atmospheric CO_2 concentrations for the past 60 years, showing a steady rise with summer/winter fluctuations. Courtesy of the Scripps Institution of Oceanography.

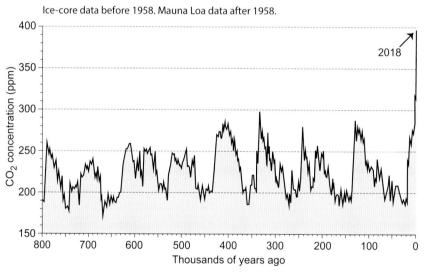

Figure 11.4 The ice data and the Keeling curve: CO_2 concentration over 800,000 years; now one third higher than any previous peak, and still climbing. Courtesy of the Scripps Institution of Oceanography.

earliest likely emergence of *Homo sapiens*. And this dramatic change has occurred very rapidly, since the dawn of the Industrial Revolution, but mostly since 1950. This telling illustration is the reason, above all, that scientific consensus has been reached on man's contribution to climate change.

The role of methane

Methane or marsh gas (CH_4), the miners' fire damp, was also shown by Tyndall to be a greenhouse gas, some 28 times more potent than CO_2. Its main sources in the atmosphere are animal stomach and intestinal gases, turnover of soil in agriculture, the cultivation of the world's most important crop in rice paddies, melting permafrost, and release by mining including the new technique of fracking. In all these circumstances it has been produced originally by bacterial decomposition of organic matter in conditions of low oxygen. In contrast to CO_2, it is broken down in the atmosphere relatively quickly into CO_2 and water by reaction with hydroxyl (OH^-) radicals. Analysis of concentrations in ice cores has shown that, in prehistory, levels were reasonably stable, but that they started to rise from around 100,000 years ago, reflecting the increasing human population and its need for agricultural produce.[17] They have now reached extraordinary levels, indicating that the rate of production of methane is outstripping the atmosphere's ability to break it down (Fig. 11.5). However, the rate of increase is variable for reasons that are not clear, and in some years there is even a fall, presenting a research challenge for atmospheric chemists.

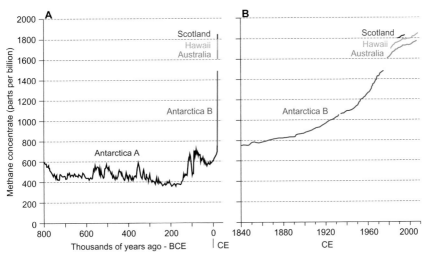

Figure 11.5 Methane concentrations in ice cores (**A**) and recent atmospheric measurements in selected countries (**B**). Note concentrations are in parts per billion, not per million as for CO_2. Courtesy of NASA.

The overall picture

The science of climate change is necessarily complex, but a clear picture has emerged and coal is only a part of the story, albeit the most important part historically. There is indisputable scientific evidence that CO_2 and methane (and to a lesser extent many other compound gases such as volatile organic compounds), act to trap heat in the planet and that concentrations of both gases are rising rapidly to levels well above any that mankind has experienced since even before we first migrated from Africa. The evidence suggests that almost half the CO_2 in the air is from fossil fuel combustion, and that much of the methane derives from man's farming and mining activities. The explanation of the rise is simply that production of these gases has been outstripping the capacity of the oceans, vegetation, and other natural processes to remove them from the atmosphere. Anyone looking at the above figures must feel some unease; we are all surely and unwittingly participating in a global experiment designed by ourselves. Unfortunately it is a badly designed experiment, since there is no control group. Scientific experiments on environmental change would subject half the subjects to the change, leaving the other half unexposed. In this case, however, the experimental subjects are not only the entire human race; the experiment is on all living organisms on the planet, and we will all respond in different ways. What might be the consequences? What can be done about it? These questions are the subject of the next two chapters.

Chapter 12

The 21st century: the world and its changing climate

The United Kingdom has undergone three major changes in my lifetime, in all of which coal played a central role. The first was from 1939 to 1950: the Second World War, the subsequent election of a Labour government under Clement Attlee, and the adoption of Keynesian economics in the introduction of the Welfare State. The health of the coal industry was central to Britain's survival through this period, leading to the mineworkers' trade union assuming a powerful role in politics. The second was our accession to the European Union in 1973 under the Heath Government, finally joining the great pan-European movement of unity, against war and for trade and security. Again, this movement started with the formation of the European Coal and Steel Community with its tripartite organisation of management, workers and governments. The third started with the election of Mrs Thatcher's government in 1980 and its adoption of President Reagan's neo-liberal economics, which looked back to Friedrich Hayek's post-Second World War advocacy of free trade, low taxes and minimal government interference. This was seen as a redressing of the balance between management and labour, widely thought to have swung too far in favour of powerful trade unions, and resulted in a crippling strike, the defeat of the miners, and the eventual closure of the British deep coal mining industry.

This third, neo-liberal, period was adopted across the Western world and by governing parties of both Left and Right, and has led to a progressive dilution by governments of their democratic responsibilities in controlling their economies. The alternative to government control is control by big business and its wealthy, often rather anonymous owners, a form of plutocracy. This, hardly noticed by the ordinary voter, led to the increasing power of massive corporations shielded from the responsibility of paying fair taxes, off-shore owners of hedge funds, irresponsible major banks that lost sight of any moral obligation to their customers, and the financial crisis of 2008. The current period of political and social turmoil, with widespread discontent and the rise of both right wing and nationalist movements is a consequence. Many are bewildered, adrift on a sea of uncertainty, wondering what is going on. But everywhere you look, coal, oil and gas, fossil fuels, are part of the problem. At the root of this lie three things: the world's population; our need for energy, food and water; and tribalism. And of all the existential threats to civilisation exercising the minds of social scientists, the greatest is now climate change and its consequences.

In his book, *Collapse: how societies choose to fail or survive*, Jared Diamond analyses the factors likely to have been responsible for the failure of (or, in some cases, its avoidance by) past societies. Common to these, lies a failure of group decision-making when confronted with a problem. This may be (1) failure to anticipate that a problem might exist, (2) failure to notice it when it arrives, (3) failure to respond to it when it has arrived, either for rational or irrational reasons, or finally (4) failure to act because solutions do not exist. Considering the history of coal, it is possible to see that for a long period the perceived benefits to society far outweighed (in the eyes of both politicians and voters) the obvious societal and health disadvantages to those who laboured to produce the resource. Few, I think, reacted to pit disasters by renouncing the use of coal. This balance shifted when the risks were perceived also to affect the whole urban population, leading in the UK and Europe to progressive legislation controlling coal use. The same is now happening with oil in response to air pollution information, and individuals and companies are slowly moving to electric-powered vehicles and equipping their houses with solar panels. Up to this point, collective and governmental responses in the West and increasingly in the East have been rational if insufficient, and have recently begun to have a modest effect. But we are now at a point when the threat from fossil fuel combustion is aimed at society on a global level in a globalised world. We are waiting, at stage 3 of the above process, to see whether a global response can be organised and will be effective.

In this chapter I shall summarise the basis for the conclusion that the Earth is retaining excess solar radiant heat, and that this is causing a change in the climate that portends very significant and disastrous effects on civilisation and biological systems. I shall discuss in my final chapter what steps we should collectively and individually take to prevent our reaching stage 4. But first, a note on the roots of controversy.

Controversy

There are two words in English that sometimes get confused: cynicism and scepticism. Both derive from ancient Greek philosophical schools and both imply disbelief. However, scepticism, doubt about accepted belief, is a necessary attitude in science without which significant advance in thinking would be almost impossible; scepticism requires a serious attempt to understand the *status quo* before attacking it. In contrast, cynicism implies contempt for accepted belief, usually attributing it to ulterior motives without troubling to understand the reasoning behind it. Scepticism is a laudable human response to information, very necessary now in an era of false news and 'tweetery'. Cynicism is a natural human response that is, in contrast, unworthy and requires to be kept in check; sometimes this is difficult.

Often the sceptics in science are ultimately proved wrong. Science and consequent technologies advance in two distinct ways: incremental small advances, as illustrated by the improvements to the steam engine in chapter 2 or the development of organic chemistry in chapter 4, or major new insights that change everything, such as the many ideas that contributed to climate change science. These new insights rightly attract scepticism, as they start as ideas that require to be tested; they may well not be based on secure experimental foundations. And if they threaten to overturn someone else's cherished ideas they may provoke enmity and even cynicism. A classic example is the response of the established nineteenth-century Christian church to the theory of evolution, echoes of which still resound among less intellectually developed societies in parts of the USA and Africa. If, however, the new idea does not gain support from evidence, it will pass into the waste bin of science and the sceptic will be justified. Sometimes the cynic may also find justification. For a long time many believed, without obvious evidence, that the tobacco industry was being disingenuous in its claim that cigarettes were not addictive and that their advertising was not aimed at recruiting young smokers. Eventually the Faustian deal between these companies and their advertisers and public relations people was exposed and the facts became apparent; the cynics were justified.

Few people bothered about the science of climate change until quite recently, and it was possible to organise a poll on the subject that would divide populations into three; those who regarded it as a threat to humanity, those who thought it a hoax, and a majority who neither knew nor cared. A group of the population became defined as 'climate change deniers'. Some of these were genuine sceptics regarding parts of the science, with valid questions about it, but the majority were simply cynics or worse, people whose wealth depended on the continued exploitation of fossil fuels and who would defend it against all rational argument. The problem was compounded by the apparent desire of the news media to present the public with a balanced view, without understanding that an even balance requires equal weight of evidence on both sides. Few deniers were familiar with the science as outlined above, but most of the argument related not to the science but to speculations about consequences. What are these supposed consequences?

The rising temperature

Readers will recall those two Scottish pioneers, Joseph Black and James Watt, remembered for their insights into chemistry and engineering, but who independently arrived at the concept of latent heat. The word 'latent' means hidden, heat that is introduced into a system but does not always lead to a rise in temperature. The energy is used in another way, such as changing the state of the heated matter from solid to liquid or liquid to gas. Watt showed how the energy could be used to drive engines more efficiently, to give motion to bodies. Black showed how heat could drive chemicals to combine or dissociate. And we now know that the sun's radiation gives us almost all the energy that sustains life on this planet (geothermal energy may sustain some deep-sea organisms). That energy not only maintains or raises the temperature; it also moves the water and evaporates it, melts the ice, powers the winds and storms in the air and the eddies in the oceans, and enables life to exist and thrive on land and in the seas. As when a kettle is heated, so when more of the Sun's energy is retained by the Earth's atmosphere, some of it remains latent and the temperature does not necessarily rise relentlessly or evenly. The pre-1940 observations of Arrhenius and Callendar nevertheless suggested a steadily rising global temperature in their time, although earlier accounts of the Earth's climate suggested that significant changes in temperature up and down over relatively short periods had occurred; examples were the warm period in the first millennium CE which allowed Europeans temporarily to colonise Greenland, and the cooler period sometimes called the Little Ice Age from about 1300 until the early twentieth century. People of my age certainly have noticed significant changes in the temperatures since the 1940s in the UK with accompanying earlier springs, warmer winters and changes in the seasons of plants and animals around us (Fig. 12.1).

Going back much earlier, the ice core data mentioned in the last chapter has shown there to have been dramatic changes of several degrees in global temperature over short periods of a few years between the low points of the ice ages. There has been much speculation as to the causes of these fluctuations, which must have had significant effects on the whole biology of the planet. Were they a result of variations in sunspot activity, changes in the tilt of the Earth, volcanic activity with massive

Figure 12.1 As I wrote, a Comma butterfly arrived in my garden in Edinburgh, the first I had seen in 40 years. These were, until recently, confined to the far south of England.

emissions of dust, or changes in the strength and direction of the great ocean currents? It seems likely that periodic fluctuation in the distance of the Earth from the Sun caused by alterations in its axis of rotation is a primary cause, but many other factors may have contributed. It is easy to see why the early investigators of climate change faced the criticism that the changes they were observing were no more than the usual cyclical variation in the Earth's climate, and nothing to do with human activity. This was especially so, as many scientists involved in studying past ice ages were warning of the possibility of another one in the future, and indeed, studying the phenomenon with an eye on finding ways of preventing it.

The measurement of the planet's temperature is extremely complex, and the interested reader is referred to Spencer Weart's excellent account of the history of the scientific developments (see chapter 11, ref 10). Changes in temperature are uneven, occurring more in the Arctic than the Antarctic, more in the northern hemisphere than the southern, and of course, differing from place to place and from time to time. However, it has recently proved possible to aggregate the worldwide recorded atmospheric and ocean surface temperatures back to 1880 into one diagram, a global land–ocean temperature index in relation to the temperatures recorded in the 1960s (Fig. 12.2).[1] This illustrates that there was a rise until the 1950s, a period

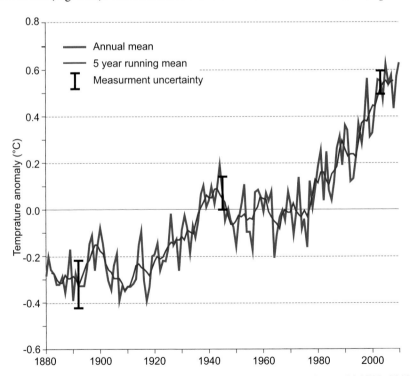

Figure 12.2 Global temperature change in sea and air across the world 1880–2010, expressed as difference from the 1960s. Data from NASA Goddard Institute for Space Studies.

of apparent stability to about 1980, and an alarming rise since then; in recent years it has been commonplace to hear that we are suffering the warmest month or year ever since records began. The cause of the period of stability is much debated, but a cooling effect of industrial pollution is now the leading explanation, and this raises the important dilemma as to whether reducing pollution might increase the warming. Another complexity is that some particulate pollutants have a warming and some a cooling effect, and this is still under investigation. But the obvious fact is that average global temperature has already risen by about 1˚C since the end of the nineteenth century.

The consequences of a rise in temperature: deaths, deserts and floods

To a doctor there is an obvious health implication of a rising temperature; more elderly people die of heart failure, and in already hot places everyone is at risk of death from heat stroke. Such episodes have been reported in Greece in recent years, and it would be very surprising if many such deaths did not occur, probably unnoticed by many of us, among the refugees suffering the terrible conditions in camps in the Middle East today. In already hot climates such as the Middle East and many parts of Africa, USA and China, water evaporation causes drought, crop failure, uncontrolled fires, and desertification. It is also obvious that in normally cold places ice will melt, and this is now well established at the North Pole (Fig. 12.3),[2] the Greenland glacier, in the West Antarctic ice shelf, and in most

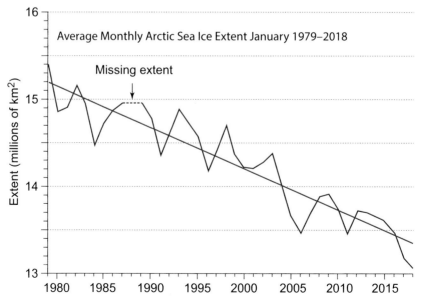

Figure 12.3 Total area of sea ice over the Arctic, showing a persistent reduction with seasonal fluctuations. Data from EPA National Snow and Ice data center.

mountain glaciers such as those in the Andes and the Alps. In the last few years it has become possible to navigate a ship across the north of Canada from Atlantic to Pacific in summer, something that was but a fatal dream to mariners such as Sir John Franklin in earlier times.

Health benefits of a warmer climate are likely to be small, since it has been easier to adapt to cold (by warm clothing and housing) than to heat. However, the distribution and timing of epidemics of infectious disease are likely to change with temperature changes. While there may have been some justification for the optimistic views of Arrhenius and Callendar that rises in temperature and CO_2 might benefit agriculture in temperate latitudes, this is probably largely offset by a less easily foreseen complication; that the water evaporated from hot regions deposits as rain in cooler places. So drought and flooding occur simultaneously in different parts of the world, both a consequence of global warming.

Sea levels

Two factors are responsible for sea-level rise – expansion of the water from a rising temperature and addition of fresh water from melting ice on land (floating ice, as on the Arctic Ocean, does not, of course, increase the volume of water when it melts).[3] The rise in sea level, now measured both by tidal gauges and by satellite, is well documented and may be taken as the most reliable index of global warming (Fig. 12.4).[4] The effects are uneven, being more significant in some places, since the land itself is not stationary, some coasts slowly rising and some sinking; this explains some divergence between the satellite and tide gauge measurements. Among the coasts sinking are those around Florida and the north of the Gulf of Mexico.

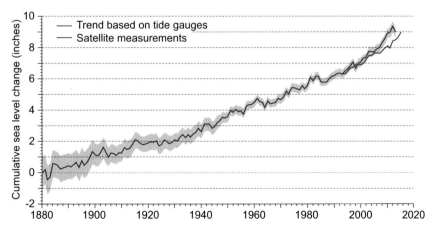

Figure 12.4 Rise of the sea level from 1880 to 2010. From the 1990s measurements by satellite (in blue) have been included. They diverge somewhat from tide gauge measurement because of land rises and falls. Data from EPA National Snow and Ice data center.

The rise of about 25cm over a century doesn't sound very much, but this has to be added to the tidal rise and the effects of storms to realise the implications. The most important effects are obviously on the most low-lying lands such as Bangladesh, the Maldives, southern Florida and Louisiana, and many small island nations. Wealthier countries such as the Netherlands and parts of USA and the UK are better able to protect themselves up to a limit, at present being seriously threatened only by storms and tidal surges, but the persistent rise means that more money will need to be spent to upgrade defences, and ultimately major cities such as London and New York will follow New Orleans and Miami in finding out the cost of climate change the hard way.

Lurking behind the hard statistics is the knowledge that vast quantities of water are presently locked in the Greenland and West Antarctic glaciers, sufficient to increase sea levels by many metres and inundate many of the world's major cities. This is one of the great imponderables, an event that may have been behind some of the past dramatic changes in climate. Assuming the present rate of increase, the International Panel on Climate Change estimates a rise of 28–98cm by the end of the twenty-first century in the absence of such major events.

Storms

Since the energy of a storm is derived from heat in the water, the warmer the water, the stronger the storm is likely to be. This explains why the most violent storms, hurricanes and typhoons, have their origin in tropical seas. As I was writing this in September 2017, a series of four violent hurricanes devastated islands in the Caribbean and northern Gulf of Mexico, going on to damage southern Texas and Florida and reminding us of the toll in human lives and suffering left behind by such disasters. While there is little evidence that such events are becoming more frequent, there is some evidence that they are becoming more violent. There remains uncertainty about the relationship between violence of storms and water temperature, since there is also some evidence that a concurrent higher air temperature may reduce the effect of the water temperature.[5] It is likely that any relationship will become clearer in the next few years.

Aside from the tragedy of deaths from injury and drowning, these storms cause huge economic problems for the communities affected and leave a legacy of homelessness and depression. Combined with sea-level rise, it would be expected that these storms will continue to cause serious problems for the insurance industry. As an example of the potential costs of sea-level rise, the flooding of New Orleans in 2005 by hurricane Katrina led to property damage costs estimated at $108 billion and the displacement of over a million inhabitants, the greatest diaspora in US history. Half never returned to their former homes. As always, the poorest suffered the most.

Feedback

Living organisms have adapted to their environments through evolution. Part of this depends on their/our ability to maintain a stable internal environment, the *milieu intérieur* of the French scientist Claude Bernard (1813–78), when our outside environment changes. When it gets too hot we lose heat by sweating; when we eat too much chocolate our rise in blood sugar is dealt with by release of insulin, and so on. In a series of notable publications, the independent scientist James Lovelock[6] proposed that the Earth itself, with its atmosphere and biology, acts like a self-regulating living organism. On the suggestion of the writer, William Golding, he called it Gaia after the Greek Earth goddess. Such self-regulation requires both positive and negative feedback systems. Unfortunately the response to warming involves a number of positive feedbacks that may well solve the problem (from the Earth system's point of view) by readjusting the biological population.

The most obvious, positive, feedbacks that amplify the effects of warming are:

- melting of ice – this not only raises water levels but also leads to less heat being reflected back into space from white surfaces, a process called albedo from the Latin for whiteness;
- release of methane from warming tundra;
- death of oceanic plankton and coral in warming and acidified tropical waters;
- death of trees and wild fires in overheated and dried-out forest;
- loss of vegetation as deserts expand;
- increase in production of CO_2 from increase in metabolism of soil micro-organisms.

Looked at this way, man's hope of a negative feedback is only likely to be realised by his reversing the way in which he has changed his environment. From Gaia's point of view, mankind may be regarded as a disease rather like the uncontrolled proliferation of cells that constitutes a cancer, which eventually destroys the organism from which it is derived.

The world's population

For most of the history of mankind since the end of the last Ice Age the estimated population of the planet barely increased; that is, the number of births and deaths were close to being balanced, a combination of disease and warfare restricting the average survival and thus life expectancy of humans. From about 4000 years ago better agriculture supported a growing population, interrupted by occasional mass epidemics such as the Black Death. However, from the early eighteenth century the growth of the population has been exponential and is now approaching eight billion (Fig. 12.5).

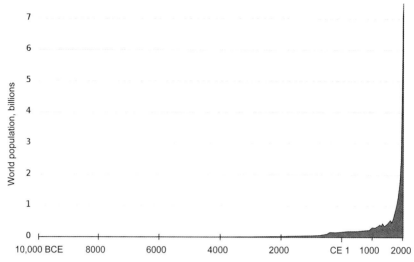

Figure 12.5 Estimated global population from 10,000 BCE. (Data from NASA.)

The date at which the population curve took off coincided with the warnings of Thomas Malthus, the pessimistic philosopher mentioned in chapter 3. The fact that his worst fears were not fully realised, and that some of his solutions seemed morally repugnant, perhaps led to his warnings being regarded as irrelevant. The ingenuity of humans allowed improvements in agriculture and better methods of distribution and storage of food, enabling the support of an increasing, and increasingly productive, workforce. The problems of periodic episodes of starvation and outbreaks of epidemic disease afflicted mostly the poor, who made up for losses with a high birth rate. This applied both in wealthier countries, but also and for much longer in the poor world, made worse by the geographic differences in the prevalence of infectious disease and the maldistribution of medical facilities. It is an unfortunate fact that an inverse care law operates, as first enunciated by Julian Tudor Hart in 1971, based on his experience as a doctor and researcher working with Archie Cochrane in the Welsh mining valleys from the 1950s; medical care is distributed in inverse relationship to the needs of populations.[7]

Tribalism and migration

To understand the implications of this population increase, it may be helpful to think in biological terms. It is plain that mankind has become the top predator, unthreatened by any species other than himself and a number of micro-organisms, perhaps rather like wolves in the wild. They have to find enough food for themselves and their offspring, and if their rivals eat too much they will be threatened, and in desperation fight with others over food. Our jungle is the planet and there are places on it where food and water are scarce, so people migrate and come into conflict with other people. The lucky and enterprising ones make their way to places where there are abundant resources, currently the overfed Western world. As water and gases flow from a place of high pressure to one of lower pressure, so animals and people migrate from places of higher to lower stress. In order to survive we all need water, energy and food. Looking at Figure 12.5, it becomes obvious that there will come a time when the population would outgrow the planet's ability to supply it with the resources required for survival. It is not simply the number of people, but also the amount we all consume. Most people aspire to become wealthier, to consume more resources, both in food and in material goods. Were the Chinese population to rise to the level of consumption of the USA, for example, that alone would double the impact of man on the environment, even with no increase in population.

In chapter 1 I touched on the origins of mankind and our migration across the planet. In the course of this migration we evolved into many different races and communities, each adapted to the local environment and shaped by it, with different physical features and means of communication. We formed tribes, then nations, then empires, and fought over our territories. Think of the history of Europe; the default position is warfare and the recent period of peace is unusual. In this we do not differ significantly from people in central Africa, the Middle East or Myanmar, or indeed, from our own country in pre-Norman conquest times when Saxons, Vikings, Britons and Picts fought for control of the island. This was appreciated by the founders of the European Union. But external pressures can break this unity and the nation states start again to feel more secure in isolation, as Shakespeare's *precious stone set in a silver sea*. As medieval barons built themselves castles to protect the territory they had gained, we talk about protecting ourselves by walls, our trade by tariffs; we withdraw into our tribes. Exacerbating this is the relentless effect of climate change and migration of people from less favoured lands. But the fact is that such withdrawal from alliances makes us weaker, more vulnerable to those very threats. Or, as Kipling put it:

> Now this is the law of the jungle –
> As old and as true as the sky;
> And the wolf that shall keep it may prosper,
> But the wolf that shall break it must die.
> As the creeper that girdles the tree-trunk
> The law runneth forward and back –
> For the strength of the pack is the wolf,
> And the strength of the wolf is the pack.

There are thus two conflicting forces at work, driven by changes in climate. As life gradually or suddenly becomes unsustainable from drought, flooding or crop failure, people migrate and have an impact on other people in their destination. These people have the choice of welcoming them or rejecting them. Some will take the former course, incorporating them into their society, while others will take the latter, emphasising religious or racial differences and forcing them to live in camps or to make further perilous journeys. In my view, although many other factors influence people's desire to move from their territory, migration is the single greatest problem that arises from or is exacerbated by climate change. It is one that is only soluble by international cooperation and treaties. The alternative is continuing starvation, disease and warfare.

Our interdependence

In the same year, 1971, that Julian Tudor Hart wrote of the inverse care law, the biologist Paul Ehrlich and physicist John Holden described the ways in which the growth in population interacted with the planetary ecology, showing the interdependence of all living things, micro-organisms, plants and animals, and their relationships to the things we require, water, food, energy and materials.[8] Changes in these relationships, such as through agriculture, industrialisation or climate variations, can have widespread and sometimes unexpected consequences elsewhere in this interdependent network, with the potential to lead to widespread environmental degradation. They concluded:

> To ignore population today because the problem is a tough one is to commit ourselves to even gloomier prospects 20 years hence, when most of the 'easy' means to reduce per capita impact on the environment will have been exhausted. The desperate and repressive measures for population control which might be contemplated then are reasons in themselves to proceed with foresight, alacrity and compassion today.

Those 20 years are long past. This and many subsequent publications on the subject constitute a warning to governments that continues to be unheeded, moving us clearly from stage 2 to stage 3 in Jared Diamond's failure sequence mentioned above. At least, since the 1970s, we have recognised the possibility of major environmental disaster. The behaviour of the human race is influencing not only the climate of the planet but also the ecology of everything that lives on it. There is now well-established evidence of migration and loss of species, for example pollinating insects, as well as humans. We have now been at the stage at which action is required for nearly 50 years. Perhaps it is too late, but we certainly need to act urgently and more effectively.

Our need for energy

This book has illustrated one very important, but not the only, aspect of man's use of energy. We have tended to think of energy resources as virtually limitless, and indeed there is plenty of coal, gas and oil still to be extracted. Figure 12.6

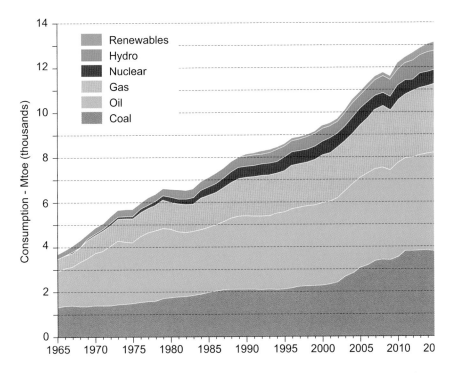

Figure 12.6 Global energy consumption 1991–2016, as megatonnes oil equivalent. Nuclear, hydro and other renewables make up only about 16% of total supply at present. (BP Statistical Energy Review 2017.)

illustrates the continuing increase in the world's demand for energy even since 1991, and the dominance of coal, gas and oil in this market. The rise reflects the increasing population but, more importantly, the requirements of developing industrial economies in China and India, both of which are still heavily dependent on cheap coal.

One sign of hope is that since 2015 there has been an overall worldwide fall in coal extraction, as China reacts to growing problems with air pollution and the knowledge of the consequences of climate change. At the same time there has also been a move towards increasing use of renewable energy in Europe, and the move from coal and oil to the somewhat less polluting natural gas in the USA. What can be achieved is illustrated by changes in Europe recently, increasing exploitation of wind and solar power. However, there is the world-wide problem of use of oil by motor vehicles and aeroplanes. It is clear that any solution must involve a complete change in our attitude to transport and renunciation of the internal combustion engine unless operated on renewable fuels.

What of the future?

Anyone reading this far must feel pretty helpless. We have seen the rise to power of King Coal and of the mighty empire that he ruled, the inventions that gave rise to prosperity for those who exploited his power in their favour. We have witnessed the rise of the usurper of his power, King Oil, and watched their tussle for power progressively damage their empire and threaten the wellbeing of its people and ecology. Like Christian in John Bunyan's 1678 allegory, *Pilgrim's Progress*, we set out with Hope but have sunk into a Slough of Despond occasioned by our own extravagance with the gift of power. Can we get out of it? What lessons can we learn from what we have seen? The next chapter discusses the international efforts being made to mitigate the effects of climate change and the contributions all of us can make. For it is something that cannot simply be left to governments: we are individually and collectively responsible and we have to deal with it.

Chapter 13

Now it is up to us

The crisis has now been recognised, and few thoughtful people doubt the fact that the climate is changing and that mankind is having a huge influence on this. The consequences are still uncertain, but are becoming better understood and are very worrying. I would liken my own feelings to those I had in the 1960s when the world approached nuclear war – a feeling of helplessness in the face of tribal aggression by the great powers. However, on that occasion enlightened diplomacy avoided an apocalypse, and there are now signs of concrete advances in both individual responses and international cooperation towards managing the consequences of climate change. There is something we can all do, collectively and individually, to avert catastrophe.

The international response

The great advances in understanding global warming in the late twentieth and early twenty-first centuries have occurred from the coming together of many different teams of investigators across the world in many different disciplines, in a joint effort to measure the changes occurring and to collate their results as part of the World Climate Research Programme, following the World Climate Conference in Geneva in 1979. The first success was the Montreal Protocol of 1987 to restrict production and use of chlorofluorocarbons, the ozone-destroying chemicals. Since then there has been an exponential growth in both the scientific understanding of the factors that determine climate and the power of computers to analyse the worldwide data collected. These computers are then used to model climates, enabling scientists to make predictions of the effects of theoretical changes in such factors as the production of CO_2 and methane, or of different atmospheric pollutants.

In 1988 the Intergovernmental Panel on Climate Change (IPCC) was established by the UN Environment Programme and the World Meteorological Organisation to consider the results of international research and report the findings. By their nature, the IPCC reports require consensus across their multi-nation participants and thus tend to the cautious side. The reports are always careful to present risks in terms of likelihood rather than certainty. The IPCC's first report in 1990 concluded that the world was indeed warming and that this was likely to get worse. By the time of its second report in 1995 it had concluded that serious warming was likely in the twenty-first century. Unfortunately the world's greatest polluter, the USA, had previously blocked action following the United Nations' Rio de Janeiro conference in 1992. By 2001 the third report of the IPCC was able to state that continued warming was very likely and that a scientific consensus had been reached. By the fourth report in 2007 it was able to show that serious climatic effects were already apparent. This collaborative work was recognised by the Nobel Prize committee awarding the Peace Prize jointly to the IPCC and to the main campaigner for action to combat the effects, ex-US Vice President Al Gore. The fifth report in 2014 went into detail in discussing the measures that required to be taken to mitigate the most disastrous consequences.[1]

The scientific evidence provided by the IPCC in 1990 led to action by the United Nations in organising the Rio Earth Summit, resulting in the production of the United Nations Framework Convention on Climate Change (UNFCC) which was ratified by the signatory countries in 1994. Its objective was *to stabilise greenhouse gas concentrations in the atmosphere at a level that would prevent dangerous anthropogenic interference with the climate system*, the signatory countries meeting annually to report progress. Scientists in the USA had taken a leading role in analysing the climate data and assessing the likely and possible consequences, but this had largely been negated by the organised opposition of vested interests, especially the coal and oil industries, and of politicians inclined or persuaded to support them. This opposition started to crumble in 1997 when John Browne, the head of the major energy company BP, recognised that action was required. In the 1997 meeting of IPCC at Kyoto, legally binding reductions in greenhouse emissions were agreed for developed countries. At Cancún in 2010, it was agreed that participating countries would act to limit global temperatures to less than 2°C above pre-industrial levels. At Paris in 2015, 195 countries signed a treaty requiring each to plan and publish its steps towards meeting the greenhouse gas targets. IPCC optimistically also lowered the target to 1.5°C above pre-industrial levels and included a requirement to report on efforts taken towards mitigating the effects of climate change.

National obligations

National governments do not readily yield to supra-national demands unless they see advantages in appropriate action. The instinct of governments is towards short-term gains in popularity in order to ensure re-election rather than unpopular moves that may antagonise voters. Thus, in spite of the best intentions expressed, and even signed up to, during international meetings, progress in achieving targets on climate change abatement has been slow. Moral argument, notably the book *A Moral Climate* by Michael Northcott,[2] Professor of Ethics at Edinburgh University, and the 2016 encyclical *Laudato si: on care for our common home*, by Pope Francis, though powerful and based firmly on scientific understanding and sympathy for those least able to protect themselves, have cut little ice with most politicians in power.[3]

What does make an impact is a strong economic argument, especially when coupled with experience of the actual adverse consequences of climate change. This was first provided by the Independent Review commissioned by the UK Government and published by Nicholas Stern in 2006. Stern, at the time a senior Treasury civil servant, wrote: *Climate change presents a unique challenge for economics: it is the greatest and widest-ranging market failure ever seen.* His views at the time were that the evidence gathered by his review led:

> to a simple conclusion: the benefits of strong, early action considerably outweigh the costs. The evidence shows that ignoring climate change will eventually damage economic growth. Our actions over the coming few decades could create risks of major disruption to economic and social activity, later in this century and in the next, on a scale similar to those associated with the great wars and the economic depression of the first half of the 20th century. And it will be difficult or impossible to reverse these changes.

If that were not strong enough, in 2016 when interviewed by *The Observer* newspaper, Stern (now a member of the House of Lords) stated that he had underestimated the risks:

> We have been too slow in acting on climate change. In particular, we have delayed the curbing of greenhouse gas emissions for far too long. When we published our review, emissions were equivalent to the pumping of 40–41bn tonnes of carbon dioxide into the atmosphere a year. Today there are around 50bn tonnes of carbon dioxide equivalent. At the same time, science is telling us that impacts of global warming – like ice sheet and glacier melting – are now happening much more quickly than we anticipated. [4]

These important statements from both moral and economic points of view coincide closely in attributing the ecological crisis that we are witnessing to a failure of the modern version of capitalism to produce a sustainable, just and equitable society, and in seeing any solutions as requiring radical change in behaviour across society. In particular, in the rich world we need to look towards a rapid move towards a less individual consumerist and a more communitarian society. We need to question the concept of economic growth as measured currently and work towards a more equitable distribution of the goods both within nations and across the planet. And clearly this can only be effective if it is coordinated on an international scale through consensus at the highest political levels.

The obvious problem facing any democratic national government is that, of necessity, their objectives are restricted by their duration in power, and this in turn is affected by their ability to please the voters, who usually also take a short-term view. Thus politicians try to please voters by pretending they can be all things to all people in the knowledge that once elected they can plead *force majeure* when confronted with the realities of power. This is why international treaties are so important. Actions to mitigate climate change do, and will increasingly, impact on the lifestyle of those of us who have become used to cheap energy, holidays, transport and goods from across the world. Politicians, in other words, have to stop promising the Earth (the metaphor seems to me prophetic as the planet is indeed being mortgaged) and address reality. And we as individuals have to stop expecting a steady progress to an imagined Utopia; the young of today bear little responsibility for climate change, yet are the ones most likely to suffer from its effects. It is on them that the future largely depends, and it is encouraging to see signs of a growing awareness of the issues among the younger generations in Europe.

Mitigation of the effects

It is noteworthy that the argument has turned from preventing climate change to accepting that the process is now well under way and the requirement is to reduce its effects. It is also clear that its progression cannot be avoided in the medium term, but that action now could reduce the rate and ultimately stop it getting worse. The Paris Agreement has obliged individual countries to set targets in terms of mitigation. These are simple to understand but undoubtedly difficult to achieve. In essence, worldwide we have been producing more greenhouse gases, especially carbon dioxide and methane, at a faster rate than they can be removed by the natural sinks – vegetation, atmospheric chemistry, and the oceans. Mitigation requires us to reduce emissions and boost the Earth's removal capacity. It is now believed from detailed modelling of the climate that stabilisation of emissions at the present level would not be effective and that reductions are essential if temperature rise is only to be limited to as much as 2°C. Thus a primary goal is to decarbonise energy, to cease the use of fossil fuels for production of electricity and for transport. It is not realistic in the medium term to move to green energy completely, so renewables need to be backed by other sources for the immediate future; at present natural gas is being used to fulfil this role, but since it is a fossil fuel, in the longer term nuclear power seems the only realistic option.

Secondly, along with finding alternative sources of power comes the need seriously to reduce our requirements for energy, by increasing the efficiency of our activities, be they industrial processes or our personal lives. Thirdly, we need to pay particular attention to the role of vegetation, forestry and farming, and how we use our land resources to feed the population. Finally, we need to grasp the nettle of population increase and aim for stabilisation.

The UK Government, in spite of some internal dissent, signed up to the Paris Agreement in 2016 but the USA Government, in an act of unbelievable short-sightedness and in contradiction to the policies of several of its more enlightened States, withdrew in 2017. This illustrates the conflict that exists in politics between doing what is necessary in the long term and appeasing one's supporters, both corporate and individual, in the short term. A series of disastrous hurricanes, floods, mud slides and wildfires in the USA shortly afterwards served to remind that nation of the risks that were being taken with their lives and livelihoods. To refer back to Lavoisier's trial, as mentioned in chapter 11, to hold that *La République n'a pas besoin de savants ni de chimistes* is a dangerous attitude for a country that is the acknowledged world leader in the scientific understanding of climate change. The much satirised presidential election mantra, 'America first', foretold the risk of a surrender by the USA of its leadership role in a globalised world and seems to have been stated in ignorance of the likely consequences.

Individual actions

There has been a marked shift in public understanding of the issues surrounding climate change. In the 1990s there was widespread ignorance; most people polled on the subject were 'don't knows'. Now, at least in the West, the majority have probably been persuaded that change is happening, and have a general appreciation of why, though only a minority are taking personal action. Nevertheless, the change has been in the right direction and will be the start of a major shift in lifestyles. It is not sufficient to recognise the problem and expect governments to do something, for the reasons given above. Each one of us must act: here are some of the things we can try to do. They all relate to lessening our impact or carbon footprint:

- When outside the home, try to use the lowest carbon method of travelling: walking, cycling, lowest carbon public transport available. If a personal car is necessary, go electric or hybrid, and drive with the lightest possible use of accelerator and brake. Always think twice before getting an aeroplane; the telephone and internet can more efficiently deal with many current excuses for air travel and trains are generally a better way for business travel within the UK, as you are able to continue working while travelling on them and don't waste time at airports. Much energy is wasted by professional people going to conferences; the same time spent reading, interacting with colleagues and searching on the internet is more efficient.

- When in the home, make sure it is well insulated and the temperature is controlled at a sensible level. Be sure to switch off all appliances when not in use and do not be profligate with unnecessary lighting. Fit solar panels for generation and for water heating if possible. Consider joining cooperative solar, wind or water power groups. One such in my area has provided all local primary schools with extensive solar panel arrays.

- Be thoughtful about your diet and limit the amount of red meat in it. The rearing of herbivorous animals for food is a very inefficient use of grain, as well as being responsible for considerable greenhouse gas emissions.

- Think what you buy. Everything in the shops has a carbon footprint. Favour local fresh produce, and avoid as far as possible plastic containers and wrapping. Plastic not only costs carbon to manufacture but also is pretty indestructible and costs more carbon to recycle or dispose of, in addition to the well-known problems associated with its dispersal through the oceans and the food chain.

- Do not be afraid to discuss the issue of climate change with your friends and colleagues. Those who understand the basics of the science can be very influential in what remains a sea of relative ignorance; spread the word.

You will notice something about this advice. In the early chapters I discussed the wonderful inventions and innovations that coal brought to our lives and how they have made life more comfortable and easier for the majority in the West. People of my generation lived through a period of austerity during and after the war that young people in the UK would find hard to believe. Now I am suggesting that we and our children need to revert to a similar, much less affluent lifestyle. It is interesting that people of my generation were generally slimmer than later generations and have lived longer than any generation before our time. The current major public health issues in the West are obesity and diabetes in middle age and the deterioration associated with old age, though you may be pleased to hear that there is some evidence that the incidence of dementia is now beginning to reduce in line with that of heart disease. All these facts probably relate to alterations in diet and exercise patterns, so the good news is that the advice given above is not only beneficial in terms of helping to reduce the effects of climate change, but is also healthy in individual terms. By helping to reduce the effects of climate change, we are also helping ourselves. And it saves us money. On the down side, there is now some evidence in the UK and USA that overall life expectancy is no longer increasing, probably attributable to the increasing maldistribution of wealth and the development of a disadvantaged underclass who lack exercise, have a poor diet and are inclined to unhealthy habits of alcohol, drug and tobacco use.

The mention of the value of exercise raises an interesting thought. Many people now go to gymnasia for exercise and even measure their energy output against electrically powered or braked machines such as treadmills and cycles. Thinking of the human- and horse-powered machines illustrated by Agricola, what a waste of human energy! Running marathons and organising Olympic Games also. Perhaps entrepreneurs might consider capturing some of it, at least to defray the electricity consumption of their establishments. Green gymnasia sounds like a good idea.

Socio-political action

The signatories to the Paris Agreement have agreed in summary to the following:

- to aim to limit global temperature increase to 1.5°C;
- to make binding commitments to nationally determined contributions (NDCs) from domestic measures;
- to report regularly on emissions and progress towards achieving their NDC, for international review;
- to submit new NDCs every five years, anticipating they will show progress;
- to adhere to obligations to support the efforts of developing countries, and encourage efforts by developing countries themselves;
- to aim at mobilising $100 million in support of these efforts through 2025, with a higher goal thereafter;
- to extend a mechanism to address loss and damage from climate change that will not provide a basis for legal liability;
- to avoid double counting in international emissions trading;
- to develop a new mechanism to enable emission reductions in one country to be counted towards another country's NDC.

These are the good intentions which the new US Administration is anxious to renounce unilaterally. It is obvious that they represent costs and actions that may be unwelcome to uninformed voters. The difficulties confronting each individual in attempting to change his or her lifestyle are as nothing compared to the problems confronting governments, since they have to persuade populations to accept unwelcome change. In order to do so, they need to be fully committed to explaining the issues and the need for that change. This is less difficult now as the evidence is becoming pretty obvious, but the sceptics and cynics still find it easy to get an equal hearing to that given by the media to climate scientists. Nevertheless, images of flooding, desertification, fires and hurricanes, and mass migration from the Middle East and central Africa are beginning to have an impact.

The action required is simple to state: governments need to decarbonise energy production and shift the emphasis of agriculture increasingly from meat to grain and vegetables. Under the Paris and earlier agreements, how this is done is left to each individual signatory and will depend on local circumstances, but plainly, simply exporting polluting activities from a rich country to a poorer one, though it may help the exporting country to reduce its footprint, does nothing to reduce the overall global issue. The use of fossil fuels must be reduced rapidly and eventually eliminated, and the loss made up as far as necessary by substitution with solar, wind, hydro, tidal and nuclear. This has to be coupled with reductions in use of energy by efficiencies, including encouragement of the personal actions suggested above, better house and workplace insulation, many more small-scale local generation initiatives, and improved electricity distribution and storage systems.

Many technological fixes, either to reduce incoming thermal radiation, to increase reflection back into space, to bury CO_2, or to increase the oceans' ability to absorb it, have been proposed, but none to date has got far beyond the drawing board. There is certainly the possibility that a temporary stay in the increasing temperature could result from massive volcanic eruption and a dust cloud in the stratosphere, but such an enforced prolonged winter would bring its own problems without providing a long-term solution. The possibility of a solution, but only so far as it applies to electricity generation in the UK, is apparent from Figure 13.1. This shows that we have been successful in substantially reducing our use of fossil fuels, renewables making an increasing contribution. However, within this coal use has almost ceased but gas use has increased. Wind and solar have increased enouragingly but, when taking account of oil use in vehicles and more generally, in 2016 renewables contributed less than 20% of our requirements. We have a long way to go, especially when transport moves almost entirely to electric power. It seems inevitable that nuclear power, whether generated locally or imported, will be required

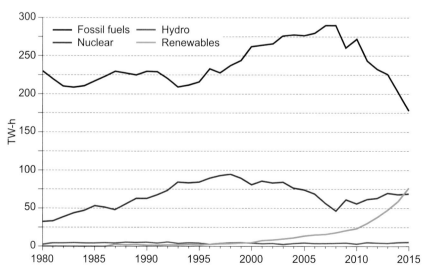

Figure 13.1 Overall reduction in use of fossil fuel and increase in renewables in Britain since 2001 (https://commons.wikimedia.org/w/index.php?curid=50666175).

indefinitely to make good the inevitable deficit in energy when renewables cannot cope.

For now, and short of a major break-through with nuclear fusion and hydrogen technology, it looks as though we have to rely on uncomfortable general contraction of our living standards in the West, while throughout the world we must require a more even distribution of resources and of wealth if the human population and our ecology are to be sustained. It is notable that in the non-capitalist world, China has formally recognised the problem of maldistribution of wealth; with total control of the economy it has embarked on an interesting path that certainly will restrict further individual liberties, especially of those who have enriched themselves, but may be beneficial for the planet and for that huge economy's further development. Arguably, with its enormous population China has most to suffer from climate change, and thus the greatest incentive and indeed opportunity to take severe measures.

The problem of a growing population hits hardest across Africa (both Kenya and Nigeria have annual growth rates of over 2.5%) and the Middle East (eg., Kuwait 4.6%, Iraq and Afghanistan c.3%). Many South American countries have now shown a slowing of growth, as have China (to 0.52%), India (to 1.26%) and the USA (to 0.7%). Nevertheless, with few exceptions population growth continues at an uncomfortable rate, worldwide at 1.18% per annum. There are no easy solutions, compulsory or regulatory methods not having proved very effective and nature's methods (starvation, conflict and disease) being abhorrent. Better methods of education, especially of women, the importance of whose education has been ignored in many countries, and increasing prosperity lead to lowering of family sizes. Some of the measures for redistributing funds in the Paris Agreement may help in this respect.

Food production and distribution

Each country needs to develop a food policy that is sustainable. It may surprise some to know that this has been necessary before in the UK and was successful. John Boyd Orr (1880–1971) was born in Ayrshire and had early experience of the effects of malnutrition when as a student in Glasgow he saw the conditions of the poor in the slums of that city. After a spell as a schoolteacher he qualified in medicine, worked briefly as a ship's surgeon then in physiology, before being appointed director of a nutrition institute in Aberdeen in 1914. Interrupted by the war, he enlisted in the Army Medical Corps and was twice decorated for gallantry in the battles of the Somme and Passchendaele. After the war he resumed his researches into nutrition of both animals and humans and raised money for what is now named, after a major donor, the Rowett Institute. He appreciated the influence food supply had had on the outcome in the later stages of the First World War and researched especially maternal and childhood nutrition. In the lead-up to the Second World War he realised that nutrition was again likely to have a decisive influence, and was able to influence the UK Government in planning a basic diet sufficient to maintain health of the population, based on the fundamental constituents: milk, eggs, bread, potatoes and vegetables.

Largely thanks to the advice of Boyd Orr, food supply during the war was controlled by rationing of the main dietary components, and we were all supplied with Ration Books containing coupons for our weekly allowances of bacon, meat, sugar, tea, lard, cheese, jam, butter and margerine, and sweets. Only one egg and one pint of milk per week was allowed, though there were concessions for children and nursing mothers. Fruit and vegetables were not restricted, but many fruits such as bananas, oranges and lemons were not available. Everyone was encouraged to grow vegetables, and land was released as allotments for this purpose. We all became familiar with the exhortation 'Dig for victory', my grandmother turning her small garden over to vegetables. Essential vitamins were provided for children and nursing mothers in the form of a daily dose of cod liver oil, the taste of which those of us who consumed it will never forget! As anticipated, the Nazi submarine blockade severely restricted the food supply to the UK, but these measures not only allowed us to survive the restrictions, but also resulted in significant increases in the health of the population by ensuring fairer distribution and much better nourishment of the poorest people. In particular, childhood mortality was substantially reduced.

Rationing was gradually wound down after the war, finally ending in 1953 (Fig. 13.2). Boyd Orr, after a period as an independent Member of Parliament and work on international committees, was rewarded by elevation to the House of Lords and a Nobel Prize for Peace. Of his many publications, a 1940 Fabian pamphlet (Fabian Tract series 251) entitled *Nutrition in War* gives an excellent short account of the essentials of a basic healthy diet that is as relevant today as it was then.

It will be some time before Western governments are forced to take such drastic measures, short of another war. Today we are in an era of what is becoming known as 'nudge economics', a means of relying on more subtle ways of changing people's habits such as by public health messages relating to prevention of obesity and its accompanying diseases.[5] Reduction of meat eating has a particularly important role, as farming livestock for meat is a major producer of methane as well as being an inefficient use of grain. In the East, rice plays a similar role, producing methane in order to grow a filling but nutrient-deficient crop, providing a challenge to a command economy. In addition, changes to the level of acidity of the seas from CO_2 absorption are interfering with the food chain by killing corals and small invertebrates, while encroachment of salt water into low-lying agricultural land such as the Mekong delta

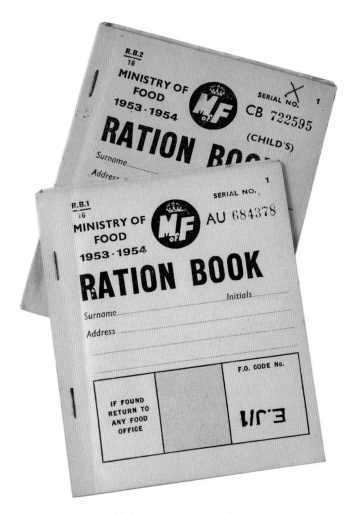

Figure 13.2 UK ration books of the 1950s.

is destroying the livelihoods of both fishermen and farmers in poor countries. Both Capitalist and Communist economic theories are challenged by climate change. The interesting and testing experiment now underway as to which system is better able to confront this, or whether there is a workable compromise based on agreed regulation, is one on which the future of civilisation may depend.

Chapter 14

There is a tide in the affairs of men

We all know what growth means; it is something we have all experienced. It starts at the beginning of life and continues for a while, then stops. Not all of it stops at the same time, but sooner or later it stops dead. It is uneven. Our brains grow rapidly *in utero* and childhood, then fall away slowly in later life. We shoot up in adolescence and shrink a bit when we become elderly. All species, plant and animal, are subject to this biological rule.

In my younger years as a doctor, we used the word 'growth' as a euphemism for cancer. It seemed brutal to tell someone outright in the clinic that they had cancer; instead we'd say that the X-ray showed a growth, and this always elicited the response 'Is it cancer, doctor?' Leaving the patient to use the dread word first seems cowardly, and changes in attitudes accompanying the general understanding that cancer could sometimes be cured allowed us eventually to come to the point more directly. It was never easy, and even after many years I still found it difficult to break the bad news. The characteristic of a cancer is that it exhibits unrestrained growth, not just in the bulk of the tumour but in spreading around the body. It is a parasite, sucking nutrition from the body until its host expires, unless it is overcome and destroyed by the body's defences or, more usually, by the power of medicine. If the cancer wins, it wins a pyrrhic victory as it dies also, no longer sustained by the blood and nutrients of its victim. All growth eventually ceases.

In the 1970s, I recall suggesting to an American colleague that the economic model on which growth is predicated was unsustainable in the long term and thus growth would one day cease, leading to financial collapse of those countries that abused it.[1] The thought had not occurred to him, and he was shocked that anyone could hold such dangerous views; indeed in those days the Earth did seem to have almost limitless resources. But I remained troubled and became more so as the years passed and we became an increasingly 'consumer society', wanting so much more than we needed; indeed, people became described as 'consumers' as though this were their defining characteristic, rather than perhaps 'contributors'. It was not possible to think of a civilisation that had not collapsed; why should ours be any different? On holiday I once visited some of the deserted Mayan cities in Mexico and thought, as I surveyed the ruins (Fig. 14.1), of the hubris of the financial centre, Canary Wharf, in

Figure 14.1 Ruins in a Mayan city, part of a great empire swallowed by the jungle after collapse of its civilisation. (Photo James Ogilvie)

London. When and in what manner would ours collapse? The photograph of the London financial centre illustrates the two threats to our civilisation – the potential for damage from rising sea levels and from air pollution (Fig. 14.2).

Figure 14.2 London's financial centre, rising through a layer of air pollution and threatened by rising sea levels. (Photo Nick Bates)

In this book I have traced the history of coal, a commodity that changed the world and was primarily responsible for the growth of its economy but which also, with other fossil fuels, has encouraged and allowed a barely sustainable growth of the population of the world and the accompanying threat of another mass extinction of species. I have mentioned Jared Diamond, the author of *Collapse*. In that book he traced the history of the decline of past populations and civilisations, identifying five important explanations: environmental damage, a change in climate, hostile neighbours, decreased support by previously friendly neighbours, and finally an inadequate response to looming problems. Even dominant societies in their day, the Greeks, Romans and the Ottomans, commercially and militarily powerful, cultured and artistic, may be numbered, along with smaller societies like the Vikings in Greenland and the Mayans, among people whose society has declined in response to these factors. Look at those factors again with respect to the increasing kleptocracy of the capitalist West and post-Communist Russia and its satellites: think of overfishing and huge agricultural monocultures, of rising global temperatures and of ships sailing to the Far East over the top of Siberia or via the North West Passage; think of the internecine strife of the Arab World and even, in a more local context, of the growing hostility of closer, previously friendly trading partners. Finally, think of the appalling delays by governments in recognising the systemic problems that face us, and of their inadequate responses to date.

All growth requires fuel, nourishment. The fuel of financial growth in society is money, but money is nothing more than pieces of paper or the flow of electrons for bartering; its value comes from what it can purchase and it is produced only by selling goods and products, be they tangible or intellectual, turbines or literature. Goods ultimately come from our endeavours, and a society that cannot produce and sell its goods shrivels and dies just as surely as does a cancer victim. Goods also require the beneficial support of our environment to produce our food, water, mineral and energy resources. Everything is a matter of balance. We can take more resources out and produce more goods, sell more than we import and have a positive balance, growth, or we can do the opposite and have a negative balance, recession. Just like a middle-aged person, we can eat more and get fat or diet and get thinner, but for most of our lives further growth is not necessary. Our society is at this stage; we do not need to continue to consume more of the planet's resources to grow, as we have more than enough. Our problem is the grossly disparate distribution of goods, whether in a small country like Scotland, in the UK, or in the wider world. This was the issue that the great nutritionist, Boyd Orr, was most concerned about – maldistribution of the necessities for life.

While I have been writing, I have had in mind those words that Shakespeare put into Brutus's mouth in *Julius Caesar*:

There is a tide in the affairs of men
Which, taken at the flood, leads on to fortune;
Omitted, all the voyage of their life
Is bound in shallows and in miseries.
On such a full sea are we now afloat,
And we must take the current when it serves,
Or lose our ventures.

I have tried to give the bad news of our condition in the understanding that, while the future looks uncertain and perilous, it is not too late to prepare ourselves to overcome the worst. All people of my age in Europe can recall times of severe austerity and many unfortunate people around the world are now suffering even worse pains than we ever did. We know that it is possible to live reasonably at a much lower level of affluence, but can we as a nation, as a community, as the fortunate minority in the world, accept it?

Jared Diamond's book has a sub-title, *How societies choose to fail or survive*. We do have a choice. We can make a fairer and more environmentally friendly world, but we need politicians who have the moral courage to grasp the nettle, and we all need to accept that our society is middle-aged and cannot keep growing forever. Necessarily, this will mean what many would regard as oppressive regulation and reduction of some individual freedom for the greater good, rather as the philosopher of Utilitarianism, Jeremy Bentham, proposed to achieve the greatest happiness of the greatest number. Come what may, the young of our country are now confronting this, but we can all still work to make it easier for them. The greatest gift we 'consumers' can give to our young people is to reduce our consumption of food, of plastics and of fuel.

The story of coal is a parable for our times and our technologies, a story of rise, abuse and decline, a story of mankind's hubris and possible nemesis. I leave you to ponder the lessons and to do what you can.

References and notes

Chapter 1

1 Fascinating and authoritative accounts of the early origins and distribution of mankind are *Sapiens: a brief history of mankind*, by Yuval Noah Harari, Vintage Books, London, 2011 and, specifically to Britain, *Homo Britannicus: the incredible story of human life in Britain*, by Chris Stringer, Penguin Books, London, 2006.

2 See Stephen Oppenheimer, *Out of Eden: the peopling of the world*, Constable and Robinson, London, 2004.

3 The main complaints came from London, to which the coal was mostly delivered by sea. A famous account, in the form of a plea to King James the II, was *Fumifugium: or the inconvenience of the aer and smoake of London dissipated*, by the diarist, John Evelyn, in 1661. This is discussed in Chapter 10.

4 Some believe that the term was derived from the fact that coal was often picked from the seashore in north-eastern England, and indeed still may be. However, its use in *Fumifugium* suggests that it was then used to describe the coal brought to London by sea. Perhaps both are true.

5 Peter Frankopan's book, *The Silk Roads: a new history of the world*, Bloomsbury, London, 2015, gives a detailed and very readable account of the importance of trade routes in determining the course of history.

6 Bryan Sykes' book, *Blood of the Isles*, Bantam Press, London, 2006 gives a fascinating account of the genetic history of the British people and their descent.

7 Latent heat is heat absorbed by a substance without alteration in temperature when it changes, for example, from liquid to gas, as in boiling a kettle. The understanding of this is relevant to climate change, as discussed in chapters 11 and 12.

Chapter 2

1 Much information on early mining in Britain is available in R.L. Galloway, *A History of mining in Great Britain*, published in 1882. It is available on Amazon (https://www.archive.org/stream/historyofcoalmin00gallrich?ref=ol#page/n11/mode/2up – accessed March 2017).

2 The translation of *De Re Metallica*, illustrated with the original woodcuts and containing notes by the Hoovers, is published by Dover Publications Inc., 1950. Herbert Hoover was himself a distinguished linguist and mining engineer before being elected President of the USA.

3 A good description of early mining and the social conditions of the time is given in Anthony Burton's *The Miners*, Futura, London, 1977

4 A very readable account of British mining development and decline is *The Rise and Fall of King Coal* by Nick Piggott, Morton's Media Group, 2016.

Chapter 3

1 *Laissez faire* means literally 'Leave alone, Let be'. This concept of leaving the markets free of regulation arose in 18th-century France and remains central to political discourse to this day. For a readable introduction to economic thought, no book is better than Robert Heilbroner's *The Worldly Philosophers*, published by Penguin Books, 1995.

2 In Edinburgh High Street, the call 'Gardyloo' was used to warn those in the street below of the imminent shower of waste – derived from the French for 'watch out for the water', *gardez á l'eau*.

3 For example, malaria simply means bad air; it was thought in southern Europe to be caused by mists from marshy land until its transmission by mosquitoes was discovered by Ronald Ross in 1897.

4 Sydenham, T., *Observationes medicae circa morborum acutorum historiam et curationem* (Medical observations on the natural history and care of acute diseases), 1676.

5 Typhus and typhoid are different diseases derived from the same word implying stupor. Typhus is caused by viruses transmitted by biting parasites such as ticks and lice, typhoid by bacteria in contaminated water. Typhoid is caused by a bacterium of the *Salmonella* species.

6 Alexander Gordon described this method of transmission in Aberdeen in 1795 and Ignaz Semmelweis described it independently in Vienna in 1827. Neither managed to persuade the conservative medical profession of their day of the value of their observations on how to prevent it.

7 *Variola* is the medical name given to smallpox, from Latin *varius*, spotted. *Vacca* is Latin for a cow, hence vaccination.

8 'Zymotic' means from fermentation, which was the then current concept of causation. Nowadays, a comparable term would be 'inflammatory'.

9 A detailed review of British mining legislation and its development before nationalisation is given in the report of the Royal Commission on Safety in Coal Mines of 1938.

10 A fascinating account of the contrasting lives of aristocratic mine owners and their employees in this period can be found in Catherine Bailey's book, *Black Diamonds: the rise and fall of an English dynasty*, Penguin Books, London, 2008.

Chapter 4

1 *A Brief History of Iron and Steel Production* by Joseph S. Spoerl gives an accessible account of these processes (http://www.anselm.edu/homepage/dbanach/h-carnegie-steel.htm).

2 Natural gas, mostly methane, is odourless whereas coal gas has a characteristic smell. In order to provide warning of a gas escape, a malodorous chemical is added to natural gas as supplied to users.

3 A comprehensive account of the origins of the chemical industry is given in Fred Aftalion's book, *A History of the International Chemical Industry*, Chemical Heritage Foundation, 2001.

4 Distillation essentially makes use of the knowledge that when a liquid mixture is heated its different chemical constituents boil off at different temperatures and may then be condensed in order to separate them. Its best known application in Scotland is in the separation of alcohol from the mash in the production of whisky.

5 The interest in curing tropical diseases at the time came from a need to protect the health of Europeans who were increasingly colonising and trading with African nations. Syphilis, as mentioned in chapter 3, was endemic in Europe and constituted a major public health problem from its effects on the brain and the cardiovascular system. These therapeutic advances were therefore of great significance.

Chapter 5

1   The names Kanawha and Monongahala are derived from the language of the Iroquois who inhabited this part of Appalachia before the white men arrived.

2   Howard B. Lee, a West Virginian attorney, gave an account of these events in *Bloodletting in Appalachia*, West Virginia University, Morgantown, 1969.

3   Many minute organisms, including some bacteria, move through water using cilia. In animals these cilia are found lining the airways and other places, such as the tubes between the ovaries and the uterus, where movement is required. They even play a role in the development of the unborn child when the primitive organs, such as the heart and liver, move to their side from the mid-line. In the 1960s electron microscope studies showed that the biomechanism of movement is the same throughout nature and can be seen, for example, in the tails of spermatozoa.

4   Inflammation is complex, but in essence it involves the recruitment of anti-bacterial blood cells (leukocytes) with capacities to digest bacteria and to neutralise them by antibodies and other chemicals that they produce. The rise in temperature also may inhibit the viability of the bacteria.

5   This description was taken originally from John Bunyan's 1680 *The Life and Death of Mr Badman*, in which he said: 'Yet the captain of all these men of death that came against him to take him away, was the consumption, for it was that that brought him down to the grave.'

6   For a history of these discoveries see F. Ryan, *Tuberculosis: the greatest story never told*. Swift publishers, Bromsgrove, 1992.

7   A micrometre (μm) is one millionth of a metre. Most bacteria are 0.5–2μm in diameter and the cells in our blood are from 5–20μm in diameter.

8   A nanometre (nm) is a thousand-millionth of a metre. Viruses come into this size range.

9   The term 'stethoscope', meaning to look into the chest, was introduced by Laënnec to describe the wooden tube he used originally to listen to the sounds, crackles and wheezes, made by breaths in diseased lungs: R.H.T. Laënnec, *Traité de l'Auscultation Médiate*. I refer to my own 4th edition, published by J.S. Chaudé, Paris, 1837.

10 An excellent and comprehensive book on the history of medicine is Roy Porter's *The Greatest Benefit to Mankind*, published by Harper Collins, London, 1997.

11 See René Dubos, *The White Plague: tuberculosis, man and society*, Rutgers University Press, New Brunswick, 1952, 2nd edition 1987 Dubos was a soil biologist who was investigating the interactions of soil organisms with his colleagues, Waksman and Schatz, who eventually discovered streptomycin, the first effective antibiotic against tuberculosis. He later became a noted philosopher of medicine.

12 See ref. 6 above.

13 This is only part of the explanation. A more complex one is that there is a mismatch in the alveoli between the supply of air and of blood, so unoxygenated blood from the right ventricle of the heart is shunted through the lungs without picking up oxygen. This results in a deficiency of oxygen in the blood as it is then circulated round the rest of the body via the left ventricle. A blue tinge to the lips and fingers comes from reduction of the normally red oxygenated blood in the tissues.

Chapter 6

1   He wrote, *Alle Dinge sind Gift, und nichts ist ohne Gift, allein die Dosis macht das ein Dinge kein Gift ist.* ('All things are poison and nothing is without poison, only the dose makes something less a poison.') In toxicology, the study of poisons and harmful effects of substances, 'dose' refers to the amount of substance taken in. In epidemiology, the equivalent term is 'exposure', a product of the amount in the air and the duration during which the exposed subjects are in the toxic atmosphere.

2 Johnstone, J. Some account of a species of Phthisis pulmonalis, peculiar to persons employed in pointing needles in the needle manufacture. *Memoirs of the Medical Society of London* 1799; 5:89–93.

3 Scrofula is a name that was given to infection of lymph glands in the neck by tuberculosis. It was once known as the King's Evil and said to be curable by the monarch's touch.

4 Alison, W.P. Observations on the pathology of scrofulous diseases with a view to their prevention. *Transactions of the Medico-Chirurgical Society of Edinburgh* 1824;1:365–438; a recent review has shown that very many stonemasons died of silicosis in the building of the 18/19th-century New Town of Edinburgh, now a World Heritage site – see Donaldson, K., Wallace, W.A., Henry, C. and Seaton, A. Death in the New Town: Edinburgh's hidden story of stonemasons' silicosis. *Journal of the Royal College of Physicians of Edinburgh* 2017; 47:375–83.

5 Peacock, T.B. On French millstone makers' phthisis. *British Foreign Medico-Chirurgical Review* 1860; 25:214–224.

6 Greenhow, E.H. Specimen of a diseased lung from a case of grinders' asthma. *Transactions of the Pathological Society of London* 1864; 16:59–60.

7 Pearson, G. On the colouring matter of the black bronchial glands and of the black spots of the lungs. *Philosophical Transactions of the Royal Society of London* 1813; 103:159–70.

8 Donaldson, K., Wallace, W.A., Elliott, T. and Henry, C. James Crawford Gregory, 19th-century Scottish physicians, and the link between occupation as a coal miner and lung disease. *Journal of the Royal College of Physicians of Edinburgh* 2017; 47:296–302.

9 Laënnec initially used a solid piece of wood to aid transmission of sound from the patient's chest to his ear, then developed this into a wooden tube, to which he gave the name stethoscope, literally 'look into the chest'. See Seaton, A. Alas poor Laënnec. *Quarterly Journal of Medicine* 2011; 104:275–77.

10 Gregory, J.C. Case of peculiar black infiltrations of the lung, resembling melanosis. *Edinburgh Medical and Surgical Review* 1831; 36:389–94. For discussion see ref. 8.

11 The lung was rediscovered by Prof. Ken Donaldson, emeritus professor of toxicology in Edinburgh University. It shows all the features of what is now called coal workers' pneumoconiosis of the complicated type, progressive massive fibrosis. See ref. 8.

12 Marshall, W. Cases of spurious melanosis of the lungs or phthisis melanotica. *The Lancet* 1834; 2: 271–74.

13 Marshall, W. Remarks on spurious melanosis of the lungs. *The Lancet* 1834; 2: 926–28.

14 Stratton, T. A case of anthracosis or black infiltration of the whole lungs. *Edinburgh Medical and Surgical Journal* 1837; 49: 490–91.

15 Craig, W. Observations on spurious melanosis. *Edinburgh Medical and Surgical Journal* 1834; 42: 330–34.

16 For a detailed account of the early medical history of CWP, the reader is referred to three papers by Dr Andrew Meiklejohn of Glasgow University in the *British Journal of Industrial Medicine* 1951; 8:127–37, 1952; 9:93–98; and 1952; 9:208–20.

17 Greenhow, E.H. *Papers relating to the sanitary state of the people of England: being the results of an inquiry into the different proportions of death in different districts in England.* General Board of Health, Her Majesty's Stationary Office, London, 1858. In spite of its title, it did include Wales!

18 A short commentary on this report in the British Medical Journal may be found at www.ncbi.nlm.nih.gov/pubmed/20763277

19 Collis, E.L. *Industrial pneumoconioses; with special reference to dust-phthisis.* Milroy Lectures to Royal College of Physicians of London, 1915.

20 Fisher, S.W. Silicosis in British coal mines. *Transactions of the Institute of Mining Engineers of London* 1935; 88:377–84 and discussion 88:384–414.

21 Cummins, S.L. Effects of coal dust upon the pneumoconiotic lung. *Journal of Pathology and Bacteriology* 1927; 30:615–19.

22 He made this remark in discussion of a paper of his: Haldane, J.S. Silicosis and coal mining. *Transactions of the Institute of Mining Engineers* 1931; 80:415–23.

Chapter 7

1 If an organ of the body is damaged by infection or by some other inflammatory process the tissue may die and form an abscess, such as a boil on the skin. In the lung, with its connection to the outside via the bronchial tubes, the damaged tissue may be coughed up and replaced by air, forming a cavity. This was characteristic of tuberculosis and usually associated with coughing up blood, haemoptysis. It also occurs in pneumoconiosis with massive fibrosis, but here the patient may cough up sputum heavily stained with coal dust, so-called melanoptysis.

2 Cummins, S.L. and Sladden, A.F.S. Coal miners' lung: an investigation into the anthracotic lungs of coal miners in South Wales. *Journal of Pathology and Bacteriology* 1930; 33:1095–1132, and Cummins, S.L. The pneumoconioses in South Wales. *Journal of Hygiene* 1936; 36:547–65.

3 Collis, E.L. The general and occupational prevalence of chronic bronchitis and its relation to other respiratory diseases. J. Indust. Hyg. 1923; 5:264–76.

4 Collis, E.L. and Gilchrist, J.C. Effects of dust upon coal trimmers. *Journal of Industrial Hygiene* 1928; 10:101–10.

5 The lymphatic vessels are effectively drainage channels that remove fluid, cells and other matter such as proteins that have accumulated in the tissues around the capillaries, transferring them via lymph nodes to the venous blood. The nodes are power houses of the immune system and are able to call anti-infective cells to destroy any bacteria that may enter them. If the system is blocked by scarring from inflammation, caused either by TB or excessive amounts of dust, further dust inhaled cannot be removed and thus accumulates in the lung itself, causing inflammation there, and thence fibrosis. See ch 6 ref. 21.

6 Gough, J. Pneumoconiosis in coal trimmers. *Journal of Pathology and Bacteriology* 1940; 1:277–85.

7 McIvor, Arthur Miners, silica and disability: the bi-national interplay between South Africa and the United Kingdom, *c*.1900–1930s. *American Journal of Industrial Medicine* 2015; 58 (S1):23–30, http://dx.doi.org/10.1002/ajim.22509

8 Hart, P. d'A. Chronic pulmonary disease in South Wales coal mines: an eye-witness account of the MRC surveys (1937–1943). *Social History of Medicine* 1998; 11:459–68.

9 He remained mentally alert and inquisitive into very old age. He periodically telephoned me when in his 80s and 90s to ask about the latest research on occupational diseases. His oral account of his life and these investigations has been recorded by the Wellcome History of Medicine Group; see ref. 12.

10 Medical Research Council 1942 Special Report Series 243; 1943 Special Report Series 244; 1945 Special Report Series 250.

11 This confusion of CWP and silicosis continued in some countries in Europe and in the USA into the 1970s, understandably since in coal miners the two conditions may co-exist and mixed patterns of lung response may be seen on microscopy.

12 The story of the PRU has been told by one of its early members, Dr John E. Cotes, *Occupational Medicine* 2000; 50:440–49 and is also the subject of a Wellcome oral history account: Population-based research in South Wales: the MRC Pneumoconiosis Research Unit and the MRC Epidemiology Unit. See vol. 13, 2002 in www.histmodbiomed.org.

13 The range of Fletcher's understanding of the issues and his sympathy for the cause is evident from the review article he published at the time: Fletcher, C.M. Pneumoconiosis of coal-miners. *British Medical Journal* 1948; i: 1015–22, and 1065–74.

14 Sir William Beveridge's Report of 1942, requested by the wartime coalition government,

set out a plan to rid the country of want, disease, ignorance, squalor and idleness. It was not very welcome to Prime Minister Churchill or most of the Conservative Party at the time, but the post-war Labour Government based the Welfare State on its recommendations.

15 Cochrane published his extraordinary autobiography as *One Man's Medicine*, BMJ Books, 1989. It is available on Amazon. In this he describes being confronted with many cases of leg oedema due to starvation (including himself), which could have been due to lack of protein or vitamin B. He obtained yeast, a source of vitamin B, from the Red Cross and added this to the meagre rations of half; those so treated improved.

16 Stewart, A. Pneumoconiosis of coal miners: a study of the disease after dust exposure has ceased. *British Journal of Industrial Medicine* 1948; 5:120–34.

17 The pathologist Chris Wagner (see later) jokingly pointed out to me that pathologically PMF was neither necessarily progressive, massive, nor predominantly fibrotic. However, it is a useful descriptor as it usually progresses, can become very large, and at least does contain some fibrous tissue (collagen) though largely it is composed of a protein called fibronectin.

18 Fletcher, C.M. and Oldham, P.D. The use of standard films in the radiological diagnosis of coalworkers' pneumoconiosis. *British Journal of Industrial Medicine* 1951; 8:138–49.

19 Until this time, interpretation of X-ray appearances was entirely subjective and varied considerably between radiologists. Discussions and trials, using representative films, led to agreement between European and US doctors on films that could be used as standards with which to compare films taken of coal miners wherever they were studied. These films, circulated by the ILO, have allowed epidemiological studies to compare prevalence of pneumoconioses all over the world. Other films were later added in order to include the different appearances in asbestos-related diseases.

20 Cochrane, A. The attack rate of progressive massive fibrosis. *British Journal of Industrial Medicine* 1962; 19:52–64.

21 Sadly, Colin McKerrow died prematurely of a rapidly progressive bone marrow cancer in 1978. John Cotes (1926–2018) left to continue his studies of lung function on industrial workers in Newcastle and is best known for his book on lung function, which became a standard text.

22 Like Cummins, Gough was probably partly right, since this could well have been a cause in some cases, but it was far from the main cause, since most people with PMF had no evidence of TB. He was thus also wrong in generalising from his observations.

23 Cochrane, A.L., Miall, W.E. Factors influencing the radiological attack rate of progressive massive fibrosis. *British Medical Journal* 1956; ii:1193–99.

24 Cochrane, A.L., Fletcher, C., Gilson, J.C. and Hugh-Jones, P. The role of periodic examination in the prevention of coalworkers' pneumoconiosis. *British Journal of Industrial Disease* 1951; 51:53–61, and Cochrane, A.L. The attack rate of progressive massive fibrosis. *British Journal of Industrial Medicine* 1962; 19:52–64.

25 Gilson, J., Hugh-Jones, P. 'Lung function in coal workers' pneumoconiosis.' MRC Special Report Series No 290, 1955.

26 See Seaton, A. 'There's none so blind as the double blind' discuss! *British Medical Journal* 2003; 326:889.

27 Anne Cockcroft has subsequently pursued a distinguished career in international health interventions in the poor world, currently from a base in McGill University in Canada.

28 Roger Seal was one of the leading clinical lung pathologists of his era, working in the NHS. He made original observations on pneumoconiosis and farmers' lung but published relatively little, preferring to spend his weekends sailing! He was said to be the model for Nogood Boyo in Dylan Thomas's *Under Milk Wood*.

29 Cockcroft, A., Wagner, J.C., Ryder, R., Seal, R.M.E., Lyons, J.P. and Andersson, N. Post-mortem study of emphysema in coalworkers and non-coalworkers. *The Lancet* 1982; 2:600–03.

30 Seal, R.M.E., Cockcroft, A., Kung, I. and Wagner, J.C. Central lymph node changes and progressive massive fibrosis in coalworkers. *Thorax* 1986; 41:531–37.

31 Cotes, J.E. The Pneumoconiosis Research Unit 1945–1985: a short history and tribute. *Occupational Medicine* 2000; 50:440–49.

Chapter 8

1   I made this tongue-in-cheek suggestion (Seaton, A. A sovereign remedy to all diseases. *Occupational Medicine* 2012; 62:365), but since then more evidence has appeared on a broad range of preventive effects of aspirin against cancer.

2   A summary of some of this research is available as Lapp, N.L., and Seaton, A. Lung mechanics in coal workers' pneumoconiosis. *Annals of the New York Academy of Sciences* 1972; 200:433–54.

3   After my appointment in 1971 I discovered that I had replaced not one but two retired physicians, and shortly afterwards a third one retired and I had to take over his role also. Nevertheless, we wrote the book and Keith Morgan's views as recorded at the time show his mind was still open on the question of coal causing emphysema – see chapter 10 in Morgan, W.K.C. and Seaton, A., *Occupational Lung Diseases*, W.B. Saunders Co., Philadelphia, 1975. Later I summarised my views in chapter 15 of Morgan, W.K.C. and Seaton, A., *Occupational Lung Diseases* 3rd edition, W.B. Saunders Co., Philadelphia, 1995.

4   A micrometre is one-millionth of a metre. The rate at which particles fall from the air in which they are suspended depends on their size, shape and density, and particle size is expressed in a dimension as if the particle were spherical and of unit density. The physics of this is complex, and the interested reader is referred to accounts by W.K.C. Morgan and J.H. Vincent in chapters 8 and 9 of Morgan, W.K.C. and Seaton, A., *Occupational Lung Diseases* 3rd edition, W.B. Saunders Co., Philadelphia, 1995.

5   The term 'complicated' was used for PMF for a time, as it was thought that the pneumoconiosis had been complicated by the addition of TB.

6   Fay, J.W.J. The National Coal Board's pneumoconiosis field research. *Nature* 1957; 180:309–10.

7   *Alma mater* is a commonly used Latin term to mean the university from which someone graduated; few now, I suspect, know its literal meaning – bountiful mother!

8   Ergonomics is the scientific study of work. At the same time that the coal industry was working to prevent lung disease, it was also very concerned about improved design of mining equipment and methods, and ergonomics made a big contribution towards this. It not only made work safer but it also facilitated productivity and was an important part of the IOM's research programme.

9   Jacobsen, M., Rae, S., Walton, W.H. and Rogan, J. New dust standards for British coal mines. *Nature* 1970; 227:445–47.

10 Coal rank, as stated previously, is a measure of the combustibility of the coal and thus its suitability for different applications. See Bennett, J.G., Dick, J.A., Kaplan, Y.S., Shand, P.A., Shennan, D.H., Thomas, D.J. and Washington, J.S. The relationship between coal rank and the prevalence of pneumoconiosis. *British Journal of Industrial Medicine* 1979; 36:206–10.

11 Miller, B.and Jacobsen, M. Dust exposure, pneumoconiosis and mortality of coal miners. *British Journal of Industrial Medicine* 1985; 42:723–33.

12 Data from www.gov.uk/statistical-data-sets/ historical-coal-data-coal-production-availability-and-consumption-1853-to-2011.

13 For commentary on this episode, see Donald MacIntyre: How the miners' strike (1984/5) changed Britain for ever. *New Statesman* 16 June 2014 (http://www.newstatesman.com/ politics/2014/06/how-miners-strike-1984-85-changed-britain-ever. Accessed March 2017.

14 Prof. A.G. Heppleston became professor of pathology in Newcastle and retired early to work

part-time in the IOM in Edinburgh, where he was able to continue his studies of pneumoconiosis for the last 15 years of his career. His key work was Heppleston, A.G., The pathological anatomy of simple pneumokoniosis of coal workers. *Journal of Pathology and Bacteriology* 1953; 66:235–46.

15 Love, R. and Miller, B.G. Longitudinal study of lung function in coal-miners. *Thorax* 1982; 37:193–97.

16 Soutar, C.A. Update on lung disease in coalminers. *British Journal of Industrial Medicine* 1978; 44:145–48.

17 Marine, W.M., Gurr, D. and Jacobsen, M. Clinically important respiratory effects of dust exposure and smoking in British coal miners. *The American Review of Respiratory Disease* 1988; 137:106–12.

18 Cockcroft, A., Wagner, J.C., Ryder, R., Seal, R.M.E., Lyons, J.P.and Andersson, N. Post-mortem study of emphysema in coalworkers and non-coalworkers. *The Lancet* 1982; 2:600–03.

19 Ruckley, V.A., Gauld, S.J., Chapman, J.S., Davis, J.M., Douglas, A.M., Fernie, J.M., Jacobsen, M.and Lamb, D. Emphysema and dust exposure in a group of coal workers. *The American Review of Respiratory Disease* 1984; 129:528–32.

20 Ken Donaldson had joined IOM from school as a technician. He was awarded an NCB scholarship and obtained a first class degree in biology at Stirling University. Back at IOM he obtained a PhD and later was appointed Professor of Pulmonary Toxicology at Edinburgh University, retiring in 2015.

21 Brown, G.M.and Donaldson, K. Inflammatory responses of lungs of rats inhaling coalmine dust: enhanced proteolysis of fibronectin by bronchoalveolar leukocytes. *British Journal of Industrial Medicine* 1989; 46:866–72, and Brown, G.M., Brown, D.M. and Donaldson, K. Inflammatory response to particles in the rat lung: secretion of acid and neutral proteinases by bronchoalveolar leukocytes. *Annals of Occupational Hygiene* 1991; 35:389–96.

22 Seaton, A., Dick, J.A., Dodgson, J. and Jacobsen, M. Quartz and pneumoconiosis in coal miners. *The Lancet* 1981; ii:1272–75.

23 Sir Richard Doll CH, FRS (1912–2005) was a very supportive member of IOM's management council. He had been Regius Professor of Medicine in Oxford and was a world-renowned epidemiologist, famous for his work on smoking and occupational and radiation-induced cancers.

24 The Colt Foundation is a charity established specifically to support research and training of young researchers in occupational and environmental medicine. From its inception in the early 1980s it has been especially helpful to the Institute of Occupational Medicine in its research and teaching activities and it is fair to say that its support was critical to the survival of the Institute after its loss of support from the coal industry.

Chapter 9

1 Seaton, A., Lamb, D., Rhind Brown, W., Sclare, G. and Middleton, W.G. Pneumoconiosis of shale miners. *Thorax* 1981; 36:412–18.

2 The early history of the oil-shale industry was researched by Dr Stephen Louw while working at IOM (published in his MD thesis for the University of Cape Town). An oral history was researched by Dr Sarah Randall. These are described in a series of technical memoranda obtainable from the IOM (TM 85/02-04 and 90/02).

3 The quotations are from the author's copy of *The Chirurgical Works of Percivall Pott*, published in 1775. The spelling and punctuation are Pott's. The Colic of Poitou (sic) was a name for lead poisoning – as with Ramazzini's name, Pott made a spelling error.

4 Bell, J. Paraffin epithelioma of the scrotum. *Edinburgh Medical Journal* 1875; 22:135–37.

5 Scott, A. On the occupational cancer of the paraffin and oil workers of the Scottish shale oil industry. *British Medical Journal* 1922; 2:1108–09.

6   Southam, A.H. and Wilson, S.R. Cancer of the scrotum: the aetiology, clinical features, and treatment of the disease. *British Medical Journal* 1922; 2:971–73.

7   Seaton, A., Louw, S.J. and Cowie, H.A. Epidemiologic studies of Scottish oil shale workers: I. Prevalence of skin disease and pneumoconiosis. *American Journal of Industrial Medicine* 1986; 9:409–421. Louw, S.J., Cowie, H.A. and Seaton, A. Epidemiologic studies of Scottish oil shale workers: II. Lung function in shale workers' pneumoconiosis. *American Journal of Industrial Medicine* 1986; 9:423–432. Miller, B.G., Cowie, H.A., Middleton, W.G. and Seaton, A. Epidemiologic studies of Scottish oil shale workers: III. Causes of death. *American Journal of Industrial Medicine* 1986; 9: 433–446.

8   Härting, F.H.and Hesse, W. Der Lungenkrebs, die Berkrankheit in den Schneeberger Gruben. *V. Gerichliche Medicin* 1879; 31:102–32. This paper is reassessed by Greenberg, M. and Selikoff, I.J. Lung cancer in the Schneeberg mines: A reappraisal of the data reported by Härting and Hesse in 1879. *Annals of Occupational Hygiene* 1993; 37:5–14.

9   Miller, B.G. and Jacobsen, M. Dust exposure, pneumoconiosis and coalminers' mortality. *British Journal of Industrial Medicine* 1985; 42:723–733.

10   Manouvrier, A. Maladies et hygiène des ouvriers travaillant à la fabrication des agglomères de huile et de brai. *Annales d'Hygiène Publique et de Médecine Légale* 1876; 45:459–82.

11   Passey, R.D. Experimental soot cancer. *British Medical Journal* 1922; ii:1112–13. Kennaway left a detailed account of his research into carcinogens in Kennaway, E. The identification of a carcinogenic compound in coal tar. *British Medical Journal* 1955; ii:749–52. An obituary is available in *British Medical Journal* 1958; i:104–6.

12   Kennaway, E. Some problems of the study of cancer in man. *British Medical Journal* 1955; i:1107–10.

13   An eye-witness account of these later years has been published by Robert Waller who worked with him and continued himself to investigate air pollution into his old age: Waller, R.E. 60 years of chemical carcinogens: Sir Ernest Kennaway in retirement. *Journal of the Royal Society of Medicine* 1994; 87:96–97. It may be of relevance to note that Kennaway would have had considerable exposure to solvents in his work, and this is now known to be a risk factor for Parkinson's disease.

14   Kennaway, N.M. and Kennaway, E.L. A study of the incidence of cancer of the lung and larynx. *The Journal of Hygiene* 1936; 36:236–37.

15   Carol Redmond summarised the available data in: Cancer mortality among coke oven workers. *Environmental Health Perspectives* 1983; 52:67–73.

16   Hurley, J.F., Archibald, R.M., Collings, P.L., Jacobsen, M., Fanning, D.M. and Steele, R.C. The mortality of coke workers in Britain. *American Journal of Industrial Medicine* 1983; 4:691–704.

17   Armstrong, B., Hutchinson, E., Unwin, J. and Fletcher, T. Lung cancer risk after exposure to polycyclic aromatic hydrocarbons: a review and meta-analysis. *Environmental Health Perspectives* 2004; 112:970–78. This estimate was a 20% increase in risk of lung cancer in relation to 100μg/per cubic metre times years of exposure. This implies that 10 years on the ovens exposed to PAH concentrations of 10μg/per cubic metre would entail a 20% increased chance of getting the disease.

18   An excellent review of benzene and its toxic effects is by Robert Snyder: Leukaemia and benzene. *International Review of Environmental Research and Public Health* 2012; 8:2975–93.

19   Case, R.A.M. Some occupational carcinogens. *Proceedings of the Royal Society of Medicine* 1969; 62:1061–66.

20   So-called acro-osteolysis, which really only translates it into Greek – painful erosions of the ends of the fingers, including skin and bone, with cold sensitivity.

21   This is a rare and aggressive malignant tumour of blood vessels, in this case in the liver. Its rarity made it relatively easy to attribute to the work when it appeared first in people exposed to vinyl chloride.

22 Lingeman, C.H. The vinyl chloride story. *The Bulletin SPEP* 1974; 4:9–15. For medical details, see Berk, P.D. Vinyl chloride-associated liver disease. *Annals of International Medicine* 1976; 84:717–31 and Simonato, L. *et al.* A collaborative study of cancer incidence and mortality among vinyl chloride workers. *Scandinavian Journal of Work, Environment and Health* 1991; 17:159–69. .

23 I am indebted to my colleague, Prof. Paul Blanc, for sending me a copy of his book (*Fake silk: the lethal history of viscose rayon*. New Haven, Yale University Press, 2016), to which interested readers are referred for a forensic account of carbon disulphide and these industries.

Chapter 10

1 Chlorofluorocarbons are also powerful greenhouse gases – see next chapter.

2 Dennekamp, M., Howarth, S., Dick, C.A.J., Cherrie, J.W., Donaldson, K. and Seaton, A. Ultrafine particles and nitrogen oxides generated by gas and electric cooking. *Occupational and Environmental Medicine* 2001; 58:511–16.

3 I investigated this and found no evidence that it caused asthma in the locality where it was grown, but it did irritate the airways of people who already had asthma. See Seaton, A. and Soutar, A. Oilseed rape and seasonal symptoms. *Clinical and Experimental Allergy* 1994; 24:1089–90.

4 In looking at the literature on air pollution, the reader will come across the terms 'particle' and 'particulate', both used as nouns, and synonymous. The old pedant in me wishes that people knew the difference between nouns and adjectives, but the realist shrugs his shoulders and accepts that when Americans change English we end up copying them.

5 The principle is the same as for the MRE113a dust sampler discussed in chapter 8, save that the orifice rather than the elutriator is designed to exclude the larger particles and admit only those below a selected size. In both cases, this allows the mass of particles below that size to be weighed. Highly sensitive microbalances are required to weigh microgram quantities.

6 The report of the committee has been translated from the Flemish and summarised by Prof. Benôit Nemery and his colleagues in: The Meuse Valley fog of 1930: an air pollution disaster. *The Lancet* 2001; 357:704–08.

7 Firket, J. Fog along the Meuse valley. *Transactions of the Faraday Society*, 1936, 32:1192–96.

8 Ammonia (NH_3), a strongly alkaline gas, was another discovery of Joseph Black in 1756. It combines with sulphuric acid in pollution episodes to form a neutral salt, ammonium sulphate ($[NH_4]_2 SO_4$).

9 McDonald, J.C., Drinker, P. and Gordon, J.E. The epidemiology and social significance of atmospheric smoke pollution. *American Journal of Medical Science* 1951; 221:325–42.

10 See Holland, W.W. and Reid, D.D. The urban factor in chronic bronchitis. *The Lancet* 1965; i:445–48. Walter Holland (1929–2018) came to Britain with his family as refugees from Hitler and was a medical student at St Thomas's hospital in London during the 1952 smog. He became one of Europe's most influential researchers in, and teachers of, public health medicine.

11 By now, everyone in the West is aware that cigarettes cause lung cancer and damage to the lungs leading to cough and breathlessness. In contrast, prior to the late 1950s most smokers actually believed that cigarettes were beneficial in helping them to cough up their phlegm. Of course, they did not realise that the cigarettes were usually the cause of the phlegm in the first place.

12 This figure is from https://uk-air.defra.gov.uk/assets/documents/reports/ cat05/0408161000_Defra_AQ_Brochure_2004_s.pdf

13 Holland, W.W. *et al.* Health effects of particulate pollution: reappraising the evidence.

American Journal of Epidemiology 1979; 110:525–659 and Shy, C.M. Epidemiological evidence and the United States air quality standard. *American Journal of Epidemiology* 1979; 110:661–71.

14 Information on changes in UK air pollution and their relevance is available on a DEFRA publication: https://uk-air.defra.gov.uk/assets/documents/reports/cat05/0408161000_Defra_AQ_Brochure_2004_s.pdf

15 Dockery, D.W. and Pope, C.A. Acute respiratory effects of particulate air pollution. *Annual Review of Public Health* 1994; 15:107–34.

16 David Bates' influential views on medical science and policy were published in 1994 in *Environmental health risks and public policy. Decision making in free societies*, by University of Washington Press, USA. See also: Bates, D.V. Health indices of the adverse effects of air pollution; the question of coherence. *Environmental Research* 1992; 59:336–49.

17 Dockery, D.W., Pope, C.A., Xu, X., Spengler, J.D., Ware, J.H., Fay, M.E., Ferris, B.G. An Association between Air Pollution and Mortality in Six U.S. Cities. *New England Journal of Medicine* 1993; 329:1753–59 doi:10.1056/NEJM199312093292401. PMID 8179653.

18 Pope, C.A. Respiratory disease associated with community air pollution in Utah, Salt Lake City and Cache Valleys. *Archives of Environmental Health* 1991; 46:90–97.

19 Schwartz, J. and Marcus, A. Mortality and air pollution in London: a time series analysis. *American Journal of Epidemiology* 1990; 131:185–94. This first drew attention to the low concentrations at which death rates were increased. An interesting aspect of this analysis was that the rate of deaths per microgram reduced at very high concentrations. This has been interpreted as being a consequence of very small particles coagulating when their concentration is very dense – see Maynard, A.D. and Maynard, R.L. A derived association between ambient aerosol surface area and excess mortality using historic time series data. *Atmospheric Environment* 2002; 36:5561–67.

20 It is interesting that the title reflects the emphasis of thought at that time on lung disease, the authors' focus being on the stronger correlation with this than with heart disease. See ref. 15 above

21 Oberdorster, G., Gelein, R., Fern, J. *et al.* Association of particulate air pollution and acute mortality: involvement of ultrafine particles? *Inhalation Toxicology* 1995; 7:111–24.

22 Although this was known in theory at the time, we later measured it as part of a study to look at the relative effects of larger and small particles on the lung. Osunsanya, T., Prescott, G. and Seaton, A. Acute respiratory effects of particles: mass or number? *Occupational and Environmental Medicine* 2001; 58:154–59.

23 I discussed this idea with three young colleagues over a pint of beer and then we wrote the paper – see: Seaton, A., MacNee, W., Donaldson, K. and Godden, D. Particulate air pollution and acute health effects. *The Lancet* 1995; 345:176-78. The central idea, that inhalation of nanoparticles might cause effects on organs away from the lung, together with the knowledge that such particles cause different effects than those of larger ones of the same constituents, was an important step towards the new science of nanotoxicology. It is nice to record that those three colleagues later all became professors themselves.

24 Early support came from Peters, A., Döring, A., Wichmann, H.-E. and Koenig, W. *The Lancet* 1997; 349:1582–87 and Peters, A., Dockery, D.W., Muller, J.E. and Mittleman, M.A. Increased particulate air pollution and the triggering of myocardial infarction. *Circulation* 2001; 103:2810–15. doi.org/10.1161/01.CIR.103.23.2810

25 Cicero in his book on the law said *Salus populi suprema lex esto* which may be translated as 'The health (or good) of the people *should be* the supreme law'. Note that he did not say that it *is* the supreme law. Lawyers and scholars, of course, are paid to argue about words, Latin ones especially. What does *salus* really mean; should it really be the *supreme* law? But for me it is a useful tag.

26 Seaton, A. and Dennekamp, M. Hypothesis: Ill health associated with low concentrations of

nitrogen dioxide – an effect of ultrafine particles? *Thorax* 2003; 58:1012–15.

27 COMEAP. The mortality effects of long-term exposure to particulate air pollution in the United Kingdom: A report by the committee on the medical effects of air pollutants. UK: Health Protection Agency; 2010.

28 Brunekreef, B. and Holgate, S.L. Air pollution and health. *The Lancet* 2002; 360:1203–42.

29 Atheroma, from the Greek for porridge, describes the appearance of fatty plaques in the inner lining of arteries. These have a tendency to get inflamed and rupture, leading to clot formation in the vessel, causing its blockage and damage to the organ supplied by restricting its blood supply.

30 Pope, C.A., Burnett, R.T. and Thun, M.J. Lung cancer, cardiopulmonary mortality and long-term exposure to fine particulate air pollution. *Journal of the American Medical Association* (*JAMA*) 2002; 287:1132–41.

31 This is intriguing, as many risk factors for Alzheimer's disease are the same as those for heart disease, leading me to think that much research on dementia may be looking in the wrong direction – at the brain cells rather than the blood vessels that supply them. There is some evidence to suggest that even air pollution might affect brain function in this way. See: Clifford, A., Lang, L., Chen, R., Anstey, K. and Seaton, A. Exposure to air pollution and cognitive functioning across the life course – a systematic literature review. *Environmental Research* 2016; 147:383–98.

32 Clancy, L., Goodman, P., Sinclair, H. and Dockery, D.W. Effect of air pollution control on death rates in Dublin, Ireland. An intervention study. *The Lancet* 2002; 360:1210–14.

33 Milojevic, A., Niedzwiedz, C.L., Pearce, J., Milner, J., MacKenzie, I.A., Doherty, R.M. and Wilkinson, P. Socio-economic and urban–rural differences in air pollution exposure and mortality burden in England. *Environmental Health* 2017; 10:104 https://doi.org/10.1186/s12940-017-0314-5

Chapter 11

1 This is an extremely simple explanation of complex phenomena. Those non-biologists curious to know more will find accounts of mitochondria, the Krebs cycle and photosynthesis in school textbooks of biology or, more easily, online.

2 In 1970 the concentration of CO_2 in the atmosphere was c.320 parts per million, or 3.2ml per 10 litres of air by volume. This gives it a partial pressure of c.0.24mmHg, far too low to prevent us getting rid of the CO_2 in our blood, which is at a pressure of c.40mmHg. Fairly obviously, gases diffuse from a place of high pressure to a place of lower pressure. Partial pressures were another discovery of Dalton: the atmospheric pressure is the sum of the partial pressures exerted by all the air's constituent gases, so that of CO_2 is a rather small contributor.

3 A good account of this period is by Marisa Linton, 2006. *Robespierre and the Terror*. http://www.historytoday.com/marisa-linton/robespierre-and-terror

4 Fourier, J. Mémoire sur les températures du globe terrestre at des espaces planétaires. *Mémoires de l'Académie (Royale) des Sciences de l'Institut (imperial) de France* 1827; 7:569–604.

5 Chemical elements comprise only one substance, such as oxygen or nitrogen. Compounds contain several different elements combined, such as carbon dioxide or nitrous oxide. Tyndall showed that while the two elemental gases in air, oxygen and nitrogen, did not absorb radiant heat, when combined as nitrous oxide they did so powerfully.

6 Tyndall, J. On the absorption and radiation of heat by gases and vapour. *Philosophical Magazine* 1861, 22:169–94 and 273–86. He published many of his lectures in two volumes of *Fragments of Science*, Longmans Green and Company, London 1879.

7 I am indebted to an article by Dr F.B. Mudge of the University of East Anglia (Mudge, F.B.

The development of the 'greenhouse' theory of global climate change from Victorian times. *Weather* 1997; 52:13–17), for information on the early history of these investigations.

8   Arrhenius, S. On the influence of the carbonic acid in the air upon the temperature on the ground. *Philosophical Magazine* 1896; 41:237–76.

9   Callendar, G.S. The artificial production of carbon dioxide and its influence on temperature. *Quarterly Journal of the Royal Metereological Society* 1938; 64:223–37.

10 The scientific story behind the concept of climate change is very complex. Fortunately there is an excellent and accessible book by Spencer R. Weart, a physicist and science historian, *The Discovery of Global Warming* Harvard University Press, Cambridge, Mass. 2008. He also provides an annually updated website on the subject at www.aip.org/history/climate.

11 Erratic stones (from the Latin *errare*, to wander) are boulders of a geological composition different from that of the area in which they are found. Study of such stones and their distribution led to the concept that the Earth had been subject to past ages of glaciation.

12 C is the chemical symbol for carbon, and the superscript refers to the atomic weight relative to hydrogen (which is ^{1}H). Isotopes are the same chemical element with differing numbers of protons in their nuclei and therefore a different atomic weight; for example, stable carbon isotopes ^{12}C and ^{13}C. ^{14}C is unstable and slowly loses its extra proton.

13 Lüthi, D., Le Floch, M., Bereiter, B. *et al*. High-resolution carbon dioxide concentration record 650,000–800,000 years before present. *Nature* 2008; 453:379–82 doi:10.1038/nature06949.

14 The most likely explanation of past ice ages is the complex periodic change in the proximity of the Earth to the Sun, owing to asymmetry of its rotation. A good explanation of this is given in Chris Stringer's book *Homo Britannicus*, Penguin, London 2005. Glaciation would then affect the biology of the planet and hence the chemistry of the atmosphere.

15 Keeling, C.D., Piper, S.C., Bacastow, R.B. *et al*. Exchanges of atmospheric CO_2 and $^{13}CO_2$ with the terrestrial biosphere and oceans from 1978 to 2000. SIO Reference Series, No. 01–06, Scripps Institution of Oceanography, San Diego, 2001.

16 The Scripps Institution has produced a video explaining the Keeling curve at https://scripps.ucsd.edu/programs/keelingcurve/2017/10/

17 See NASA Earth Observatory. A global view of methane, 2016. https://earthobservatory.nasa.gov/IOTD/view.php?id=87681

Chapter 12

1   NASA Goddard Institute for Space Studies, updated from Hansen, J.E. et al. 2001: A closer look at United States and global surface temperature change. *Journal of Geophysical Research;* 106:23947–23963, doi:10.1029/2001JD000354.

2   http://nsidc.org/arcticseaicenews/category/analysis/

3   Archimedes' principle – a floating body displaces its own weight of water, the weight being proportional to the volume.

4   Commonwealth Scientific and Industrial Research Organisation 2015 update to data originally published in: Church, J.A., and N.J. White. Sea-level rise from the late 19th to the early 21st century. *Surveys in Geophysics* 2011; 32:585–602. www.cmar.csiro.au/sealevel/sl_data_cmar.html, and· National Oceanic and Atmospheric Administration 2016. Laboratory for Satellite Altimetry: Sea level rise. (www.star.nesdis.noaa.gov/sod/lsa/SeaLevelRise/LSA_SLR_timeseries_global.php Accessed June 2016)

5   Elsner, J.B., Tsonis, A.A. and Jagger, T.H. High frequency variability in hurricane power dissipation and its relationship to global temperature. *Bulletin of the American Meteorological Society* 2006 doi:10.1175/BAMS-87-6-763.

6   Lovelock's electron capture detector, invented in 1958, revolutionised the measurement of environmental chemicals and proved its value in the detection of the Antarctic ozone hole.

See his autobiography, *Homage to Gaia*, Oxford University Press, 2000. His latest book, *A Rough Ride to the Future*, published by Penguin Books in 2014 when he was 95, speculates on a future for mankind after the crisis caused by climate change.

7 Tudor Hart, J. The inverse care law. *The Lancet* 1971; 297:405–12.

8 Ehrlich, P.R., Holden, J.P. Impact of population growth. *Science* 1971; 171:1212–17.

Chapter 13

1 Climate Change 2014: Synthesis Report. Contribution of Working Groups I, II and III to the Fifth Assessment Report of the Intergovernmental Panel on Climate Change [Core Writing Team, R.K. Pachauri and L.A. Meyer (eds)]. IPCC, Geneva, Switzerland, 2014; 151pp.

2 Northcott, M.S. *A Moral Climate*. Darton, Longman and Todd, London, 2007.

3 *Laudato si, mi Signore* – 'Praise be to you, my Lord who sustains and governs us through our sister, Mother Earth, who sustains and governs us and provides us with various fruits and coloured flowers and herbs...' quotes a hymn attributed to St Francis of Assisi.

4 Report in Observer newspaper. www.theguardian.com/environment/2016/nov/06/nicholas-stern-climate-change-review-10-years-on-interview-decisive-years-humanity

5 You will have realised that this book is a part of my small effort to nudge people towards a more sustainable lifestyle.

Chapter 14

1 I was unaware then that Maynard Keynes was quoted as having said, 'In the long run we are all dead', which is, of course, true but of little comfort.

Glossary

A

aerosol: a suspension of solid or liquid particles in air. Solid aerosols may be called a dust cloud.

alveoli: the smallest air spaces of the lung where exchange of oxygen and carbon dioxide to and from the blood occurs.

amnion: the membrane within an egg that surrounds a fluid-filled cavity protecting the developing embryo.

anthracite: a hard coal of high carbon content.

anthracosis: an obsolete term for coal worker's pneumoconiosis.

anticyclone: a period of high atmospheric pressure surrounded (in the northern hemisphere) by a slow-moving clockwise air current. In winter it brings cold, still air conducive to fog and pollution episodes; in summer it brings heat waves.

asphyxiant: of gases, suffocating.

atheroma: from the Greek for porridge, describes the appearance of fatty plaques in the inner lining of arteries. These have a tendency to get inflamed and rupture, leading to clot formation in the vessel, blocking it and thus damaging the organ supplied by restricting its blood supply.

atom: a single chemical element, for example hydrogen or carbon.

autopsy: an examination of a body after death, commonly called *post mortem*.

B

BCE: before Common or Christian Era, previously BC.

bituminous: containing tar-like matter.

bronchi: the branching tubes leading from the windpipe to the furthest parts of the lung where they divide into bronchioles.

bronchioles: the smallest airways, ending in multiple air spaces, the alveoli.

C

CE: Common or Christian Era, previously AD (Anno Domini).

chronic bronchitis: a condition characterised by regular cough and production of sputum, usually associated with cigarette smoking.

cohort: in epidemiology, a group of people, usually a sample from a population, followed forward in time, particularly to determine the incidence of ill-health.

confidence interval: in statistics, the range within which the true value is likely to fall, usually expressed as a 95% probability.

consumption: a term applied originally to any disease characterised by loss of weight and general malaise, usually associated with early death. After the discovery of bacteria it was applied principally to tuberculosis.

continental drift: the theory introduced by Alfred Wegener in 1912 that there was once a large single continental landmass (Pangaea) that split into the multiple components that represent today's continents. At first scorned, the theory became accepted as evidence accumulated (primarily from the fossil record), giving rise to the theory of plate tectonics.

COPD: chronic obstructive pulmonary disease, the modern term for chronic bronchitis combined with emphysema.

CWP: coal workers' pneumoconiosis.

E

ecology: the study of living organisms in relation to their environment.

embryo: the earliest stage of development from a fertilised egg into a living organism.

emphysema: enlargement of the air spaces of the lung by breakdown of their walls, leading to inefficient gas exchange and mechanical difficulty in breathing.

epidemiology: the study of the distribution and determinants of health and disease in populations.

erratic stones: rocks of different mineral composition from the place in which they are found, having been transported by glaciers from their original situation.

exposure: in epidemiology, the product of the concentration of a toxic substance in the air and the duration during which the person is in that atmosphere, usually expressed as milligrams per cubic metre x hours.

F

feV$_1$ or Forced Expired Volume in one second: the amount exhaled in the first second of a forced vital capacity manoeuvre.

G

glaciation (ice ages): over the past 2.6 million years the Earth's surface has been partly covered by ice, and this whole period is referred to as the Quaternary Ice Age. Within this period, the extent of the ice has fluctuated and when this has been extensive it is referred to as a glacial period or a glaciation. Popularly these shorter periods are often also referred to as ice ages, the last of which occurred from about 21,000 until 11,000 years ago.

greenhouse gas: any gas that traps heat energy in the Earth's atmosphere by allowing entry to ultraviolet but absorbing part of the reflected infrared. This is a characteristic of compound gases such as carbon dioxide and methane but not of elemental gases such as oxygen and nitrogen. The analogy of a greenhouse (Treibhaus in German) was probably first introduced by von Czerny in 1881.

H

hooker: in rugby, the player in the front row of the scrum who is responsible for directing the ball with his foot to his team mates behind him.

I

incidence: in epidemiology, the rate of occurrence of the condition of interest in a population over a specified time period, e.g., 20 episodes of illness per thousand people per year.

IOM: Edinburgh Institute of Occupational Medicine.

IPCC: Intergovernmental Panel on Climate Change.

isotope: a chemical element which can exist in two or more forms owing to differing numbers of protons in its nucleus and therefore has different atomic weights, for example oxygen isotopes ^{16}O and ^{17}O.

K

kleptocracy: rule by theft. I refer to the behaviour of the super-rich who strip out the assets of states and influence governments and elections, sometimes achieving high office themselves, by the use of their wealth.

knapping: the process of striking flint with a stone or hammer to shape it into tools such as spears and arrow heads.

L

latent heat: heat absorbed by a substance without changing its temperature when it changes, for example, from liquid to gas, as in boiling a kettle.

lymphatic system: a system of fine tubes that run alongside veins, draining fluid (lymph) and cells that accumulate in tissues to collections of nodes (lymph glands) that act as filters and immune defences, and thence into major veins. It has an important role in removing inhaled bacteria and dust from the lung, but is also present in all organs save the substance of the brain.

M

macrophage: a motile defence cell in the lung and other organs that can ingest bacteria and particles and can secrete inflammatory chemicals.

mass spectrometer: an instrument that passes substances through a magnetic field and separates atoms according to their atomic weight, allowing, for example, the separation of different isotopes.

miasma: a noxious vapour in the air, originally thought to be a cause of much disease that was later shown to be infective in origin.

microgram (μg): one millionth of a gram.

micrometre or micron (μm): one millionth of a metre.

millstone grit: coarse-grained sandstones, the name being derived from its earlier use as grinding wheels in grain mills. When used for grinding knives it was responsible for much lung disease (silicosis) in workers. Specifically, in Britain 'millstone grit' is used to refer to the coarse conglomeratic Carboniferous sandstone found in Scotland and northern England, known as the Namurian (roughly middle Carboniferous in age).

molecule: a combination of atoms, for instance, hydrogen and oxygen combined to form water.

MRC: UK Medical Research Council.

mucous membrane: the internal lining of the nose, throat, bowels and related glands, in which some of the cells secrete sticky mucus and antibacterial antibodies.

N

naphtha: liquid petroleum distilled from coal tar.

nanogram (ng): one thousand millionth of a gram.

nanometre (nm): one thousand millionth of a metre.

NCB: British National Coal Board, later British Coal.

Neolithic: the New Stone Age, the period from when mankind started farming about 10,000 BCE until the Bronze Age, about 3000 BCE.

P

phthisis: an obsolete word for tuberculosis, sometimes applied to other wasting diseases including pneumoconioses.

plate tectonics: the theory that the surface layers of the Earth, divided into seven major and multiple smaller plates, move over a lower molten layer in the mantle. Where one meets another, volcanic and earthquake activity occur.

PM: particulate matter. PM_{10} and $PM_{2.5}$ define particles less than 10 and 2.5 micrometres in aerodynamic diameter, usually weighed in micrograms/cubic metre.

PMF: progressive massive fibrosis, the accumulation of large masses of dust-related inflammation in the lungs of workers with pneumoconiosis.

pneumoconiosis: any chronic (persistent) lung disease caused by inhalation of dust, usually implying the development of lung fibrosis.

pox: lesion on the skin as in chicken pox. Two specific infective diseases, the often fatal smallpox and the great pox, syphilis, which caused a chronic disease eventually leading to madness or a death from blood vessel disease, were commonly called 'The Pox'.

prevalence: in epidemiology, the frequency of the condition of interest in a population at any one time, usually expressed as a percentage.

PRU: the MRC's Pneumoconiosis Research Unit in Cardiff.

Q

quartz: the most common crystalline form of silicon dioxide.

R

rank: for coal, a measure of its combustibility or carbon content.

renewable fuels: fuel produced from renewable sources as a substitute for fossil and nuclear fuel, for example alcohols from crops, hydrogen generated by renewable energy (wind, solar, tidal) that make a smaller contribution to the carbon cycle and may reduce CO_2 production.

respirable dust: the fraction of particles in an aerosol that is small enough (generally less than about 7 micrometres in diameter) to be inhaled to the alveoli of the lung.

S

sandblasting: the process of polishing the surface of a material such as metal or denim by a blast of fine sand from a hose.

shale: a sedimentary rock including clays, quartz, limestone and organic matter, which in some deposits is exploited for oil or gas production.

silica: another name for silicon dioxide, including the non-crystalline amorphous form.

silicates: silica combined with oxygen and various metallic elements such as magnesium, calcium and aluminium. They include minerals such as mica and kaolinite, commonly known as clays.

silicosis: a progressive lung disease characterised by nodular fibrosis caused by inhalation of quartz.

smog: a blend of smoke and fog.

sputum: mucus coughed up from the airways of the lung, commonly called phlegm. It contains cells and antibodies. With infection it goes green from a pigmented enzyme, verdoperoxidase, derived from inflammatory cells.

T

tectonics: relating to the crust of the Earth, especially the surface plates floating on molten magma, the movement and interaction of which causes earthquakes.

toxicology: the study of poisons and other potentially harmful substances, including therapeutic drugs.

V

vital capacity (VC): the volume of air exhalable in one expiration after a full inhalation. If maximum effort is used, it is called the forced vital capacity (FVC).

Index

Index